# 9

SERIES IN THEORETICAL AND APPLIED MECHANICS
Edited by R.K.T. Hsieh

**SERIES IN THEORETICAL AND APPLIED MECHANICS**
Editor: R. K. T. Hsieh

---

*Published*

Volume 1: Nonlinear Electromechanical Effects and Applications
by G. A. Maugin

Volume 2: Lattice Dynamical Foundations of Continuum Theories
by A. Askar

Volume 3: Heat and Mass Transfer in MHD Flows
by E. Blūms, Yu. Mikhailov, R. Ozols

Volume 5: Inelastic Mesomechanics
by V. Kafka

*Forthcoming*

Volume 4: Mechanics of Continuous Media
by L. Sedov

Volume 6: Design Technology of Fusion Reactors
by M. Akiyama

Volume 8: Mechanics of Porous & Fractured Media
by V. N. Nikolaevskii

Volume 10: Fragments of the Theory of Anistropic Shells
by S. A. Ambartsumian

Six Lectures on Fundamentals and Methods

# Aspects of Non-Equilibrium Thermodynamics

**W. Muschik**

World Scientific
*Singapore • New Jersey • London • Hong Kong*

*Author*

**W. Muschik**

*Institut für Theoretische Physik*
*Technische Universität Berlin*
*D-1 Berlin 12*
*F R Germany*

*Series Editor-in-Chief*

**R. K. T. Hsieh**

*Department of Mechanics, Royal Institute of Technology*
*S-10044 Stockholm, Sweden*

*Published by*

World Scientific Publishing Co. Pte. Ltd.,
P O Box 128, Farrer Road, Singapore 9128
*USA office:* 687 Hartwell Street, Teaneck, NJ 07666
*UK office:* 73 Lynton Mead, Totteridge, London N20 8DH

Library of Congress Cataloging-in-Publication Data is available.

**ASPECTS OF NON-EQUILIBRIUM THERMODYNAMICS**



**ISBN** 0218-0235
**ISBN** 981-02-0087-0

Printed in Singapore by JBW Printers and Binders Pte. Ltd.

# Preface

I would like to thank warmly Professors D.R. Axelrad and R. Baliga for their kind invitation to deliver this special seminar — lectures in "non-equilibrium thermodynamics" at the Department of Mechanical Engineering, McGill University during the summer semester 1988. I was pleased with the outstanding level of discussions that took place during these lectures. In particular, I wish to express my gratitude to my colleagues Prof. Axelrad, Prof. Eu, Department of Chemistry, McGill University, Prof. Grmela, Ecole Polytechnique de Montréal, and Dr. Bhattacharya, Department of Chemistry, McGill University. Professor Axelrad has kindly undertaken to edit the manuscript.

W. Muschik

Institut für Theoretische Physik
Technische Universität PN7-1
D-1 Berlin 12
West Germany

# Contents

# Chapter 1

# Introduction

Thermodynamics is concerned with the general structure of Schottky systems [1]. By definition these systems exchange heat, work and material with their environment. Their states are described by state variables which are elements of a suitable state space. It is useful to distinguish between considering only one Schottky system in an equilibrium environment, a so-called discrete system, and between dealing with a set of sufficiently small Schottky systems each denoted by its position and time exchanging heat, work and material with its adjacent Schottky systems. In doing so we get a so-called field formulation of thermodynamics which needs different mathematical tools with regard to the description of discrete systems. But the general thermodynamical fundamentals are the same in both cases although they may appear in a different shape. Traditionally these fundamentals of thermodynamics are formulated by "Laws" which are enumerated from zero to three. Here these laws are reformulated, extended, and adapted to the requirements of modern thermodynamics: The zeroth law becomes a statement of the state space of equilibrium systems, the first law is formulated as usual, but the unique gauge of the internal energy will be proved, the second law is extended also to negative absolute temperatures, and the third is out of consideration.

Thermodynamical theories can be divided into two classes, the *probabilistic* and the *deterministic theories* (Figure 1.1). The probabilistic theories themselves are decomposed into *stochastic*, *statistical*, and *transporttheoretical* branches which are operating with totally different probabilistic concepts. The deterministic theories are split up into those which describe *discrete systems* and others which deal with *continuumtheoretical concepts*.

*Stochastic* thermodynamics [2,3] is characterized by a measure space $[\Omega, \mathcal{A}, \mathcal{P}]$ — a Kolmogorov probability algebra — which is defined on the state space of a microdomain. These microdomains form mesodomains to which the probabilistic description of the microdomains is transferred by introducing suitable mean values. The site of the mesodomain is identified with the material coordinate so getting the connection between stochastic and deterministic thermodynamics.

*Statistical* theories are marked by a distribution function or by a density operator $\rho$ which may depend on a relevant set of observables (so called Beobachtungsebene) [4] $\mathcal{G}^1, \ldots, \mathcal{G}^k$. Here we use such a description for the foundation of non-equilibrium contact quantities, such as contact temperature or non-equilibrium chemical potentials.

*Transporttheoretical* methods also use a distribution function $f(\underline{p}, \underline{q}, t)$ which is defined in contrast to the distribution functions of the statistical methods on the $(2f + 1)$-dimensional $\mu$-space of the single molecule having $f$ degrees of freedom [5,6].

In contrast to the probabilistic theories the *deterministic* or *phenomenological* theories do not take into consideration the molecular structure of materials whereas probabilistic theories even embrace this structure by microscopic models using master equations, molecular dynamics or density operators. The basic concept of phenomenological theories is that of the *macroscopic variable*. These quantities describe the state of the system which can be retraced immediately or mediately to measuring quantities of the system. Examples are volume, pressure, temperature, mass density, charge density, magnetization, pressure tensor, internal energy, etc.

As mentioned above deterministic theories are divided into those which describe discrete systems, these are Schottky systems, and those which apply continuumthe-

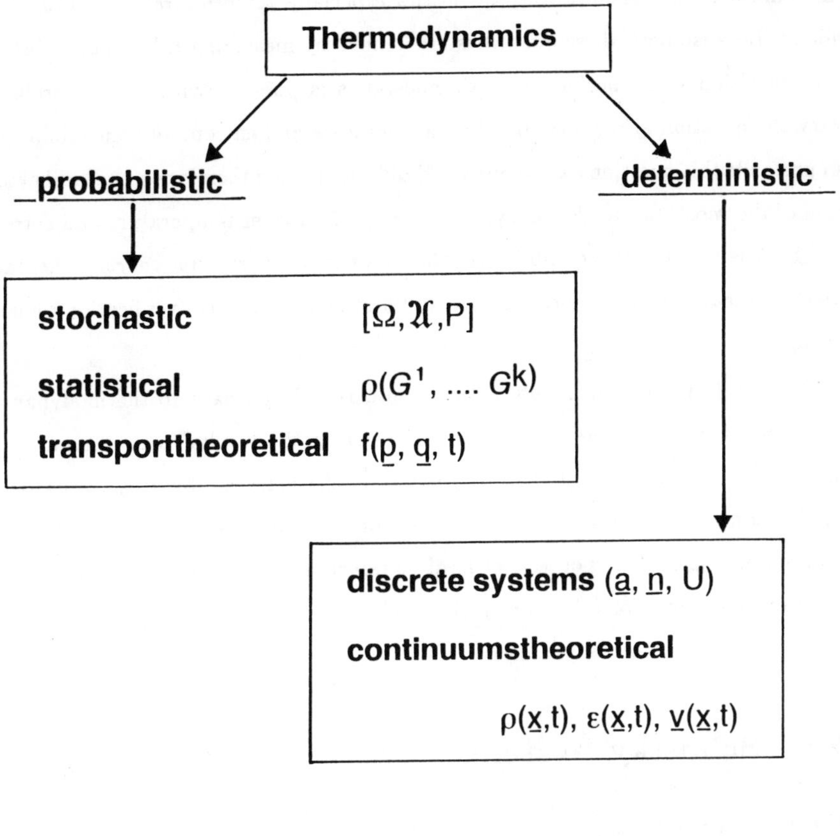


Figure 1.1: Diagram showing distinct classes of thermodynamical theories.

oretical methods. As Figure 1.2 shows there exist a lot of similar but different deterministic thermodynamical theories. It should be shortly motivated why we have such a variety of phenomenological nonequilibrium theories: The transition from mechanics to thermostatics is achieved by adding thermodynamical quantities to the mechanical ones. Besides other quantities especially *temperature* and *entropy* are added. Because both these quantities are defined by measuring rules in equilibrium, the transition from mechanics to thermostatics is possible without any problem. Now the question arises how to define temperature and entropy in nonequilibrium? In principle this question can be answered differently, and therefore no natural extension of thermostatics to thermodynamics exists [7]. Either temperature and entropy will be redefined for nonequilibrium or they are taken for *primitive concepts*, i.e. their mathematical existence is presupposed and first of all a physical verification remains open.

The six lectures deal mainly with a non-classical approach to thermodynamics starting out with discrete systems and transfering the results to continuum thermodynamics. *Non-classical* thermodynamics is characterized by a dynamical nonequilibrium concept of temperature. Therefore this approach is different from all others which use the hypothesis of local equilibrium (irreversible thermodynamics, theories using evolution criteria or variational principles) or temperature as primitive concept (rational thermodynamics).

## 1.1 Schottky Systems

### 1.1.1 Definitions

A system $\mathcal{G}$ which is separated by a partition $\partial\mathcal{G}$ from its surroundings $\overline{\mathcal{G}}$ is called a *Schottky system* [1], if the interaction between $\mathcal{G}$ and $\overline{\mathcal{G}}$ can be described by the heat exchange $\dot{Q}$, the power exchange $\dot{W}$, and the material exchange $\underline{\dot{n}}^e$ (Figure 1.3). Here the material exchange is determined by the stoichiometric equations

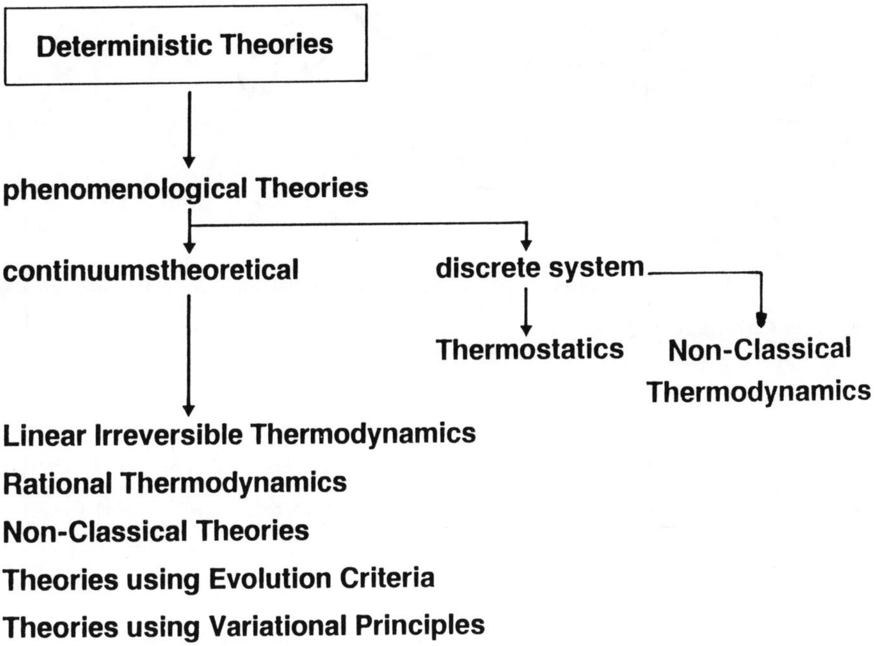


Figure 1.2: Family of deterministic thermodynamical theories.

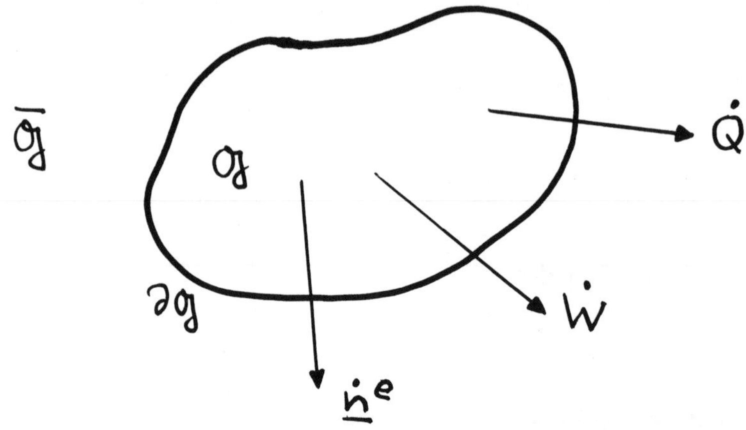


Figure 1.3: A Schottky system $\mathcal{G}$ exchanges heat, power and material through $\partial\mathcal{G}$ with its surroundings $\overline{\mathcal{G}}$.

$$\underline{\underline{\nu}} \cdot \underline{M} = \underline{0} \tag{1.1}$$

which describe the conservation of mass ($[\nu_k^r]$ = mol, $[M_k]$ = g mol$^{-1}$, $r$ = reaction index, $k$ = component index). Introducing the reaction rates $\underline{\dot{\xi}}([\dot{\xi}^r] = s^{-1})$ we get

$$\underline{\dot{\xi}} \cdot \underline{\underline{\nu}} \cdot \underline{M} = 0 \ . \tag{1.2}$$

Consequently the time rates of mol numbers due to chemical reactions are

$$\underline{\dot{n}}^i := \underline{\dot{\xi}} \cdot \underline{\underline{\nu}} \tag{1.3}$$

($[\dot{n}_k^i]$ = mol s$^{-1}$). Then the material heat exchange is defined by

$$\underline{\dot{n}}^e := \underline{\dot{n}} - \underline{\dot{n}}^i = \underline{\dot{n}} - \underline{\dot{\xi}} \cdot \underline{\underline{\nu}} \ . \tag{1.4}$$

$\underline{\dot{n}}$ is the total time rate of the mol numbers of $\mathcal{G}$.

The power exchange $\dot{W}$ between $\mathcal{G}$ and $\overline{\mathcal{G}}$ is defined by the net working

$$\dot{W} := L - \dot{E}^{\text{kin}} \ . \tag{1.5}$$

Here $L$ is the power of the forces acting on $\mathcal{G}$ and $E^{\text{kin}}$ the kinetic energy of $\mathcal{G}$ in the chosen frame.

*Problem 1.1* : Presupposing the balance equations of continuum thermodynamics there exist work variables $\underline{a}$ and generalized forces $\underline{A}$ so that the net working is a Pfaffian

$$\dot{W} = \underline{A} \cdot \underline{\dot{a}} \ . \tag{1.6}$$

Material and power exchange are chemical or mechanical quantities, whereas the heat exchange is a thermodynamical quantity which we have to define now. Therefore we use the fact that different partitions $\partial\mathcal{G}$ exist. We define a *power-insulating* and a *material-insulating* partition by the following scheme:

| partition $\partial\mathcal{G}$ is called | if holds for arbitrary surroundings $\overline{\mathcal{G}}$ |
|---|---|
| power-insulating : | $\dot{W} = 0$ |
| material-insulating : | $\underline{\dot{n}}^e = \underline{0}$ |

We now can (operationally) define what an isolating partition is: In Figure 1.4 a thermocouple TC is measuring changes of $\overline{\mathcal{G}}$ after contacting $\mathcal{G}$ and $\overline{\mathcal{G}}$.

*Def.*: A power- and material-insulating partition $\partial\mathcal{G}$ is called *isolating*, if after contacting $\mathcal{G}$ with arbitrary $\overline{\mathcal{G}}$ there is no change of the angle of deflection of the thermocouple TC. $\dot{W} = 0$ and $\underline{\dot{n}}^e = 0$ are valid by presupposition, $\dot{Q} = 0$ is valid because there is no alteration of $\overline{\mathcal{G}}$. Two further definitions are obvious:

*Def.*: If $\partial\mathcal{G}$ becomes isolating by additional power insulation, it is called *adiabatic*.

$$\underline{\dot{n}}^e = 0 \ , \qquad \dot{Q} = 0$$

*Def.*: If $\partial\mathcal{G}$ allows material exchange, $\underline{\dot{n}}^e \neq 0$, $\mathcal{G}$ is called *open system*.

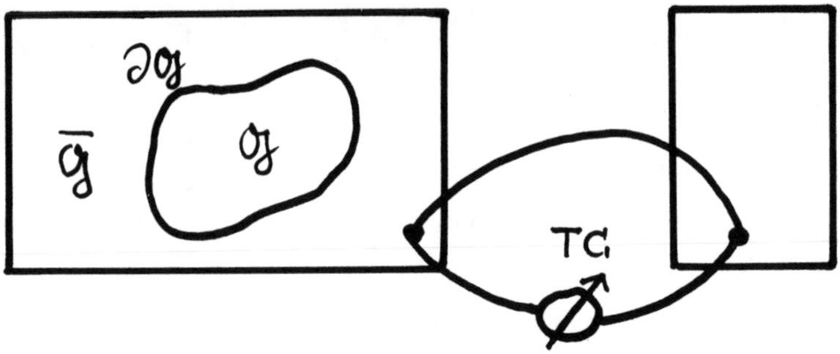


Figure 1.4: Measuring device for defining an isolation partition $\partial \mathcal{G}$.

## 1.2 State Space

For describing a state of a discrete system we need a state space. The chemical composition of the system is determined by the mole numbers $\underline{n}$, mechanical and geometrical properties are described by the work variables $\underline{a}$. We get a general state by adding thermodynamical variables $z$

$$Z := (\underline{a}, \underline{n}, \underline{z}) \in \mathcal{Z} \ . \tag{1.7}$$

$\mathcal{Z}$ is called (non-equilibrium) *state space of a discrete system* [8].

Equilibrium is defined as follows:

*Def.*: Time-independent states of isolated systems are called *states of equilibrium*.

*Problem 1.2* : Why is a stationary heat conduction process no state of equilibrium?

*Def.*: A system is called *thermal homogeneous*, if it does not contain any adiabatic partitions in its interior.

Obviously for describing a state of equilibrium we do not need as many variables as in non-equilibrium. Therefore the question arises how many variables span the equilibrium subspace? The answer is given by a law cited in the introduction: The

Zeroth Law which bases on experience and which is here formulated as an "empirem" (axiom whose validity is founded on experience).

*Empirem* : ($0^{th}$ Law): The state space of a thermal homogeneous system in equilibrium is

$$Z^* := (\underline{a}, \underline{n}, *) \ . \tag{1.8}$$

$*$ is exact one thermodynamical variable.

As we will see below this thermodynamical variable $*$ may be the internal energy $U$ of the system or its thermostatical temperature $T$.

If the state of a system changes in time, we say the system undergoes a process:

*Def.*: A path in state space is called a *process*

$$Z(\cdot) := (\underline{a}, \underline{n}, \underline{z})(\cdot)$$

especially

$$Z(t) = (\underline{a}, \underline{n}, \underline{z})(t), t \in [t_1, t_2] \ .$$

Experience shows that equal processes in different systems can induce different values of other variables not included in the state space. A very simple example are two gaseous systems having equal volume and temperature but different pressure. Of course each of the two systems consists of another gas. Different materials differ from each other by different material properties which are described by a map $\mathcal{M}$ called the constitutive map. The domain of $\mathcal{M}$ depends on the chosen state space: There are state spaces so that $\mathcal{M}$ is local in time as in the theory of thermoelastic materials or in theories using internal variables. There are other state spaces on which $\mathcal{M}$ is not local in time as in materials showing after-effects or hysteresis. Therefore we classify [9]:

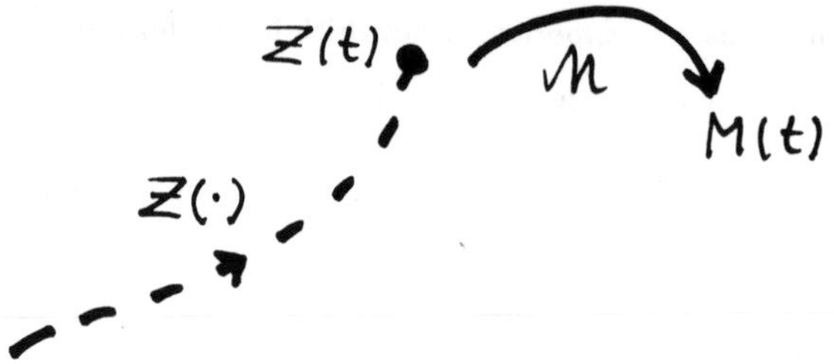


Figure 1.5: The material property $M(t)$ is only determined by the state $Z(t)$.

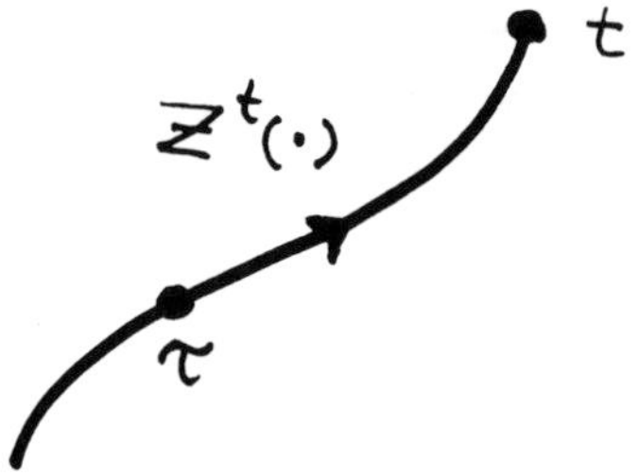


Figure 1.6: The history $Z^t(\cdot)$ of the process $Z(\cdot)$ is defined by the path of $Z(\cdot)$ up to time $t$.

*Def.*: A *state space* is called *large*, if material properties $M$ are defined by maps local in time (Figure 1.5)

$$\mathcal{M} : Z(t) \to M(t) \quad , \text{ for all } t \ . \tag{1.9}$$

For defining after-effects we need the concept of the history:

*Def.*: For a fixed time $t$ and real $s \geq 0$

$$Z^t(s) := Z(t-s) \quad , \qquad s \in [0, \tau] \ . \tag{1.10}$$

is called the *history of the process* $Z(\cdot)$ between $t - \tau$ and $t$ (Figure 1.6).

*Def.*: A *state space* is called *small*, if material properties $M$ are defined by maps on process histories (Figure 1.7)

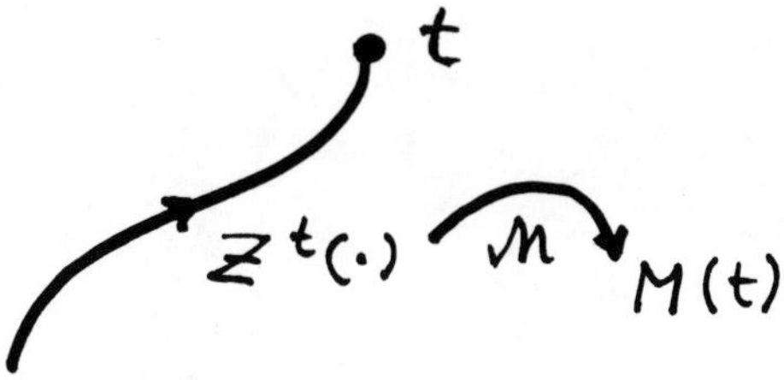


Figure 1.7: In small state spaces the domain of the constitutive map is the history of the process.

$$\mathcal{M} : Z^t(\cdot) \to M(t) \quad , \text{ for all } t \ . \tag{1.11}$$

According to the zeroth law the space of the states of equilibrium has less dimensions as the general state space. Therefore projections exist mapping the general state space onto the equilibrium subspace. By these projections to each non-equilibrium process an equilibrium "process" or better a trajectory in equilibrium subspace is attached (Figure 1.8).

*Def.*: A trajectory $Z^*(\cdot)$ in the equilibrium subspace induced by a projection $P^*$, $(P^*)^2 = P^*$, of a process $Z(\cdot)$

$$P^* Z(t) = Z^*(t) \tag{1.12}$$

$$P^*(\underline{a}, \underline{n}, \underline{z})(\cdot) = (\underline{a}, \underline{n}, *)(\cdot) \tag{1.13}$$

is called an *accompanying process*.

Because in equilibrium material properties do not depend on process histories (that is a demand for constructing the state space)

$$M^*(t) = \mathcal{M}(Z^*(t)) \tag{1.14}$$

we get by the definition of $M^*$ due to (1.11)

$$M^*(t) = \mathcal{M}(Z^{*t}(\cdot)) \tag{1.15}$$

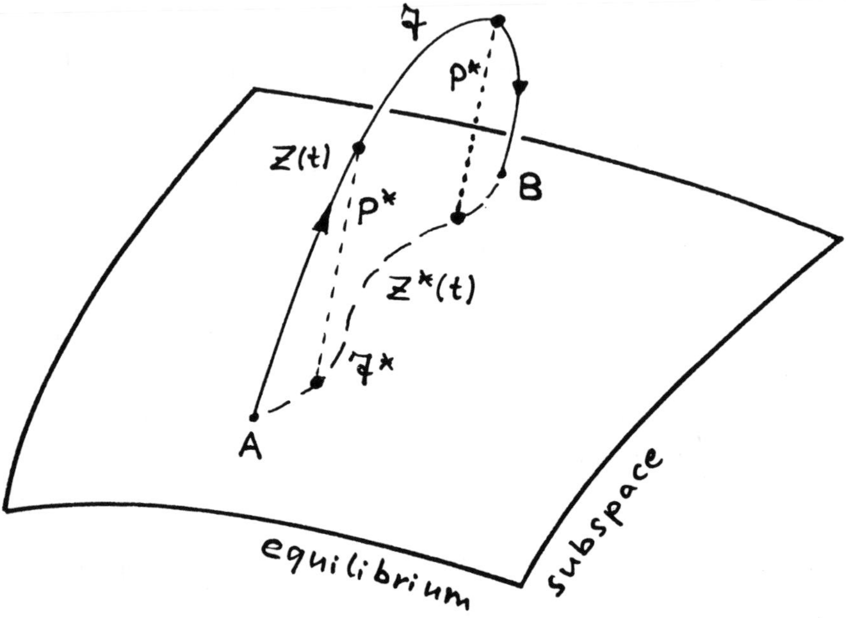


Figure 1.8: Equilibrium subspace represented as hypersurface in state space. The projection $P^*$ maps $Z(t)$ into $Z^*(t)$ so inducing the accompanying process $\mathcal{F}^*$ from the real process $\mathcal{F}$.

the following property of the constitutive map

$$\mathcal{M}(Z^{*t}(\cdot)) = \mathcal{M}(Z^{*}(t)) \ . \tag{1.16}$$

Additionally the equilibrium values $M^*$ of $\mathcal{M}$ have to be compatible with its non-equilibrium values $M$ by satisfying the

*Embedding Axiom* :

$$\mathcal{F}\int_A^B M(t)dt = M_B^{eq} - M_A^{eq} = \mathcal{F}^* \int_A^B M^*(t)dt \tag{1.17}$$

$$A, B \in \text{ equil. subsp.}$$

If we especially apply this embedding axiom to a quantity which is defined on equilibrium subspace, and we look for an extension of this equilibrium quantity to non-equilibrium, this extension has to satisfy the embedding axiom, because otherwise it would not be compatible with the earlier defined equilibrium quantity: e.g. a non-equilibrium entropy — in whatever way defined — has to obey the embedding axiom in order to be in agreement with the well known concept of equilibrium entropy.

# Chapter 2

# First Law

## 2.1 Preliminary Considerations

Figure 2.1 shows a device for measuring the dependence of the pressure on the volume for adiabatic processes. By experience we know that the special map describing the pressure as a function of volume depends on the process speed. Therefore there exist different adiabatic processes starting out of the same equilibrium state. Of course the final equilibrium states of these different adiabatic processes are different, too. Therefore it is possible to connect two states by different adiabatic processes (Figure 2.2). If we presuppose state 1 and 2 are adiabatically connected, and there are two different adiabatic processes

$$\begin{aligned} &\text{ad.: } 1 \rightarrow 3 \rightarrow 2 \\ &\text{ad.: } 1 \rightarrow 4 \rightarrow 2 \end{aligned} \tag{2.1}$$

connecting 1 and 2, experience shows that the work is independent of the path in adiabatical isolated systems:

$$W_{13} + W_{32} = W_{14} + W_{42} \tag{2.2}$$

Adiabatic processes establish a transition relation

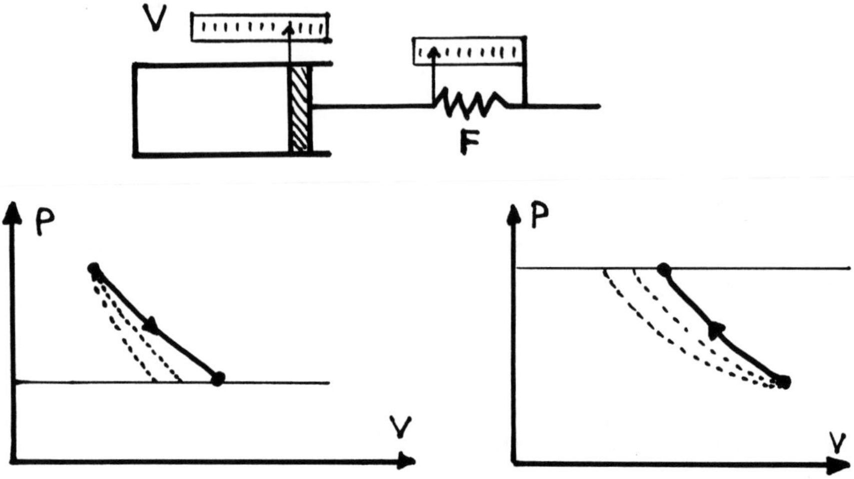


Figure 2.1: The connection between pressure $P$ and volume $V$ depends on process speed (dotted curves belong to different speeds).

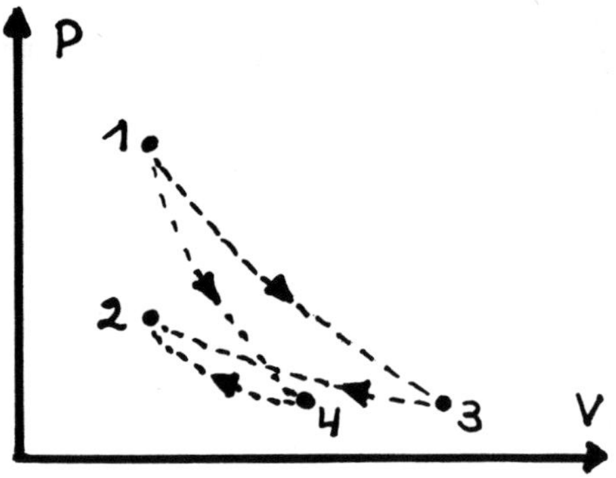


Figure 2.2: Two states 1 and 2 can be connected adiabatically by different processes.

$$Z_1 \stackrel{\text{ad}}{\to} Z_2 \tag{2.3}$$

which is *transitive*

$$Z_1 \stackrel{\text{ad}}{\to} Z_2 \stackrel{\text{ad}}{\to} Z_3 \Rightarrow Z_1 \stackrel{\text{ad}}{\to} Z_3 \ , \tag{2.4}$$

and *reflexive*

$$Z \stackrel{\text{ad}}{\to} Z \text{ (identity)} \ . \tag{2.5}$$

Defining

$$Z_1 \sim Z_2 :\Leftrightarrow Z_1 \stackrel{\text{ad}}{\to} Z_2 \vee Z_2 \stackrel{\text{ad}}{\to} Z_1 \tag{2.6}$$

we get the relation $\sim$ of being adiabatically connected.

*Problem 2.1.* : The relation $\sim$ of adiabatic accessibility is symmetric, reflexive, and transitive. Therefore it establishes a division of the states into classes of equivalence.

This division into classes of equivalence plays an essential role for proving that a unique gauge of internal energy exists.

## 2.2 Internal Energy

The independence of the adiabatical work (2.2) can be formulated as

*Empirem* : ($1^{st}$ Law): A function of state $U(Z)$ exists, so that for adiabatic processes $\alpha$

$$Z_1 \stackrel{\text{ad.}}{\to} Z_2$$

the work $W_{12}^{\alpha}$ in a system at rest is represented by

$$W_{12}^{\alpha} = U_2 - U_1 \ , \qquad U_j := U(Z_j) \tag{2.7}$$

$$\to \ \text{ad: } \dot{W} = \dot{U} \ , \ \text{restsystem} \tag{2.8}$$

If the process constraint of being adiabatical is cancelled, we define:

*Def.*: In closed restsystems the heat exchange is defined by

$$\dot{Q} := \dot{U} - \dot{W} \tag{2.9}$$

$$\rightarrow \text{ closed: } Q_{12}^{\alpha} = U_2 - U_1 - W_{12}^{\alpha} \text{ , restsystem} \tag{2.10}$$

*Problem 2.2*: If $E$ is the total energy $E := U + E_{\text{kin}} + Y$, $Y$ = potential energy, and $L'$ the power of the forces without a potential, the first law for closed systems writes

$$\text{closed: } Q_{12}^{\alpha} + L_{12}^{\prime\alpha} = E_2 - E_1 \tag{2.11}$$

## 2.3 Gauge of Internal Energy

We consider a thermal homogeneous closed system $\dot{\underline{n}}^e = \underline{0}$ without chemical reactions $\dot{\underline{n}}^i = \underline{0}$ at rest, and we introduce a *state of reference* for this system which is a state of equilibrium by definition

$$Z_o = (\underline{a}_o, \underline{n}_o, U_o) \ . \tag{2.12}$$

Now the question arises how to determine the internal energy of an arbitrary state $Z$ of the considered system taking the reference state as a basis [10].

First of all we have the

*Empirem*: Each isolated system transforms itself after a sufficient long time into a state of equilibrium:

$$1 \overset{\text{isol.}}{\rightarrow} 2^* \ . \tag{2.13}$$

Because according to the definition in isolated systems $\dot{Q} = 0$, $\dot{\underline{n}}^e = \underline{0}$, and $\dot{W} = 0$ are valid we get from (2.9)

$$U_1 = U_2^* \ , \tag{2.14}$$

and therefore to each state

$$Z = (\underline{a}, \underline{n}, \underline{z}) \tag{2.15}$$

of a thermal homogeneous system belongs an equilibrium state

$$Z^* = (\underline{a}, \underline{n}, U^*) \tag{2.16}$$

having the same internal energy

$$U^* = U(Z) \ . \tag{2.17}$$

Consequently the internal energy of non-equilibrium states can always be retraced to the internal energy of equilibrium states. So our question now runs as follows: Does a unique gauge of internal energy exist for all equilibrium states of a thermal homogeneous system?

If $Z^*$ is adiabatically connected with $Z_o$, $Z^* \sim Z_o$, we get from (2.10)

$$U^* = U_o + W_{o*}^{\text{ad}} \ . \tag{2.18}$$

The problem now arises whether all equilibrium states are adiabatically connected with $Z_o$. If it would so, a unique gauge would be achieved by (2.18). But there is no hint for the adiabatic connection of the equilibrium subspace, and therefore we have to assume that there are states $Z^*$ which are not adiabatically connected with $Z_o$ (Figure 2.3).

We now introduce two other classes of processes:

*Def.*: A process $\alpha$ is called *pure thermic*, if along $\alpha$

$$\dot{W} = 0 \ , \quad \underline{\dot{n}}^e = \underline{0} \ , \quad \underline{\dot{n}} = \underline{0} \tag{2.19}$$

are valid.

According to (2.10) we get for pure thermic processes $\alpha$

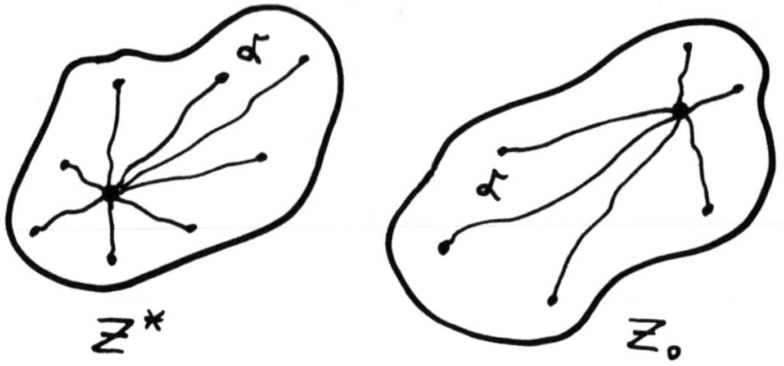


Figure 2.3: The reference state $Z_o$ is not adiabatically connected with $Z^*$.

$$Q_{12}^{\alpha} = U_2 - U_1 \ , \tag{2.20}$$

and therefore

$$Z_o \overset{\text{Pth}}{\rightarrow} \hat{Z} : \hat{U} = U_o + Q_{o\wedge}^{\text{Pth}} \ . \tag{2.21}$$

The second class of processes is:

*Def.*: A process $\alpha$, $Z' \overset{\alpha}{\rightarrow} Z''$, is called *work-cyclic*, if

$$\underline{a}' = \underline{a}'' \ , \tag{2.22}$$

and if along $\alpha$

$$\dot{Q} = 0 \ , \qquad \underline{\dot{n}}^e = \underline{0} \ , \qquad \underline{\dot{n}} = \underline{0} \tag{2.23}$$

is valid.

According to (2.10) we get for *work-cyclic* processes $\alpha$

$$W_{12}^{\alpha} = U_2 - U_1 \ , \tag{2.24}$$

and therefore

$$Z_o \overset{\text{WC}}{\rightarrow} \tilde{Z} : \tilde{U} = U_o + W_{o\sim}^{\text{WC}} \ . \tag{2.25}$$

We now introduce the following sets:

$$Z^* \in \ \text{Ad}(Z_o) :\Leftrightarrow Z^* \sim Z_o \ , \tag{2.26}$$

$$Z^* \in C_y(Z_o) :\Leftrightarrow \underline{a}(Z^*) = \underline{a}(Z_o) \wedge (Z^* \in \ \text{Ad}(Z_o)) \ , \tag{2.27}$$

$$Z^* \in \ \text{Pth}(Z_o) :\Leftrightarrow \underline{a}(Z^*) = \underline{a}(Z_o) \wedge \underline{\dot{n}}^e = \underline{0} \wedge \underline{\dot{n}} = \underline{0} \ . \tag{2.28}$$

According to the definitions we get immediately

$$C_y(Z_o) \subset \text{Ad}(Z_o) \ , \tag{2.29}$$

$$C_y(Z_o) \subset \text{Pth}(Z_o) \ , \tag{2.30}$$

$$[\text{Ad}(Z_o)/C_y(Z_o)] \cap \text{Pth}(Z_o) = \emptyset \ . \tag{2.31}$$

*Problem 2.3*: From (2.26) and (2.27) follows

$$\text{Ad}(Z_o) \cap \text{Pth}(Z_o) = C_y(Z_o) \ . \tag{2.32}$$

We consider a $\tilde{Z}$ different from $Z_o$ having the property

$$\tilde{Z} \in C_y(Z_o) \rightarrow \tilde{Z} \in \text{Pth}(Z_o) \ . \tag{2.33}$$

According to (2.21) and (2.25) the internal energy of $\tilde{Z}$ is

$$\tilde{U} = U_o + Q_{o\sim}^{\text{Pth}} \ , \tag{2.34}$$

$$\tilde{U} = U_o + W_{o\sim}^{\text{WC}} \ . \tag{2.35}$$

Consequently

$$Q_{o\sim}^{\text{Pth}} = W_{o\sim}^{\text{WC}} \tag{2.36}$$

follows. This relation means that heat exchanges of transitions between states of $C_y(Z_o)$ can be mechanically measured. Consequently we constructed a ***restricted calorimeter***, "restricted" because arbitrary heat exchanges have to be measured in more than one step. Introducing restricted calorimeters we achieved in principle the possibility to measure an arbitrary heat exchange mechanically.

It is now easy to prove the existence of a unique gauge of internal energy. Because we deal with equilibrium states of thermal homogeneous closed systems without chemical reactions an arbitrary state is represented by $(\underline{a}, U)$ according to the zeroth law. We now consider the process

$$(\underline{a}_o, U_o) \overset{\mathrm{ad}}{\rightarrow} (\underline{a}, U_1) \overset{\mathrm{Pth}}{\rightarrow} (\underline{a}, U) \ . \tag{2.37}$$

We have for the partial processes

$$U_1 = U_o + W_{01}^{\mathrm{ad}} \ , \tag{2.38}$$

$$U = U_1 + Q^{\mathrm{Pth}} \ . \tag{2.39}$$

Because $Q^{\mathrm{Pth}}$ is measurable by a restricted calorimeter $U$ is determined. Consequently a unique gauge of the internal energy exists for all states of equilibrium, and as stated above also for the non-equilibrium states.

## 2.4 Open Systems

The first law for closed systems (2.9) is

$$0 = \dot{U} - \dot{Q} - \dot{W} \ . \tag{2.40}$$

If the partition $\partial\mathcal{G}$ is now permeable for material, the left hand side of (2.40) is no longer zero, but homogeneous in $\underline{\dot{n}}^e$

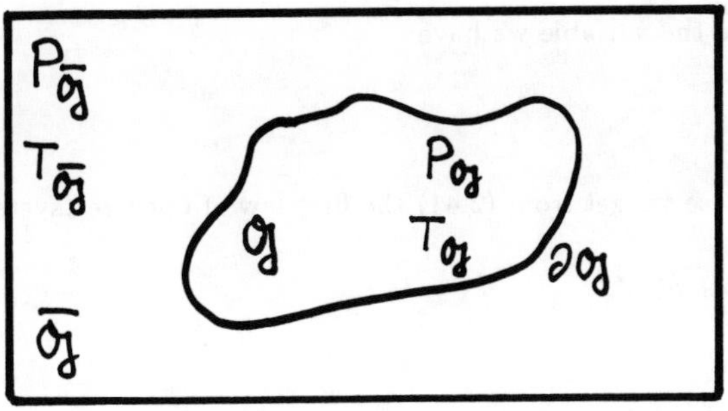

Figure 2.4: A 1-component equilibrium system $\mathcal{G}$ in a vicinity of equal pressure, equal thermostatic temperature, and equal composition.

$$\underline{X} \cdot \dot{\underline{n}}^e := \dot{U} - \dot{Q} - \dot{W} \tag{2.41}$$

For elucidating the meaning of $\underline{X}$ we especially consider a 1-component equilibrium system $\mathcal{G}$ surrounded by a vicinity $\overline{\mathcal{G}}$ of equal composition, thermostatic temperature, and pressure (Figure 2.4). If now the partition $\partial\mathcal{G}$ is removed, volume and mole number of $\mathcal{G}$ changes by constant pressure, temperature, and vanishing heat exchange:

$$\dot{p} = 0 \ , \quad \dot{Q} = 0 \ . \tag{2.42}$$

We get

$$\dot{W} = -p\dot{V} = -(pV)^\bullet \ , \tag{2.43}$$

and

$$\dot{U} - \dot{W} = \underline{X} \cdot \dot{\underline{n}}^e = (U - pV)^\bullet \ . \tag{2.44}$$

This yields

$$\dot{Q} = 0 \ , \quad \dot{p} = 0 : \underline{X} \cdot \dot{\underline{n}}^e = \dot{H} \tag{2.45}$$

Because $\dot{n}^e$ is here the variable we have

$$X \equiv \frac{\partial H}{\partial n} \ , \tag{2.46}$$

and the general case we get from (2.41) the first law of open restsystems

$$\dot{U} = \dot{Q} + \dot{W} + \underline{h} \cdot \underline{\dot{n}}^e \ ,$$
$$\underline{h} := \frac{\partial H}{\partial \underline{n}} \equiv \underline{X} \ . \tag{2.47}$$

# Chapter 3

# Second Law

## 3.1 Thermostatic Temperature

We consider the set of all thermal homogeneous equilibrium systems $A, B, \ldots$ and the thermal contact ($\underline{\dot{n}}^e = \underline{0}$, $\underline{\dot{a}} = \underline{0}$) between two of them.

By this thermal contact the relation $\sim$ of ***thermal equilibrium*** is defined by

$$A \sim B :\Leftrightarrow \dot{Q}(A, B) = 0 \ . \tag{3.1}$$

By experience we get the following

*Empirem*: The relation $\sim$ of thermal equilibrium defined on thermal homogeneous systems is
reflexive: $A \sim A$,
symmetric: $A \sim B \rightarrow B \sim A$,
transitive: $A \sim B \wedge B \sim C \rightarrow A \sim C$.

Therefore $\sim$ establishes a division into classes of equivalence. Each of these classes is denoted by a quantity $\theta$ which we call empirical ***temperature***:

$$A \sim B \rightarrow \theta_A = \theta_B \ , \tag{3.2}$$

$$A \not\sim B \rightarrow \dot{Q}(A,B) \begin{cases} > 0 \rightarrow \theta_A < \theta_B \ , \\ < 0 \rightarrow \theta_A > \theta_B \ . \end{cases}$$

The empirical temperature of the ideal gas $T$, $pV = nRT$, is called ***absolute temperature***, which will be treated in more detail below.

## 3.2 Negative Absolute Temperature

We consider a system of $N$ particles with spin $(1/2)\hbar$ in an exterior magnetic field $\underline{B}$ (sketch). If $n$ spins are parallel to $\underline{B}$, the number of states is

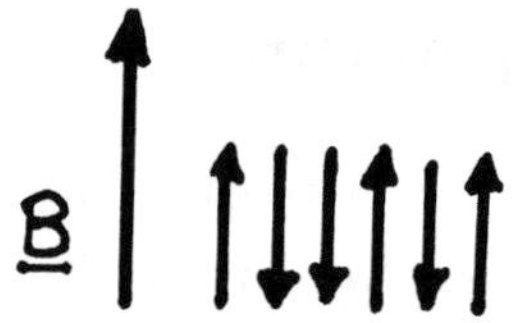


$$F_N(n) = \frac{N!}{n!(N-n)!} \tag{3.3}$$

If e.g. $N = 6$ the following table shows the number of states, their internal energy $U(n)$, and their thermostatic temperature $T(n)$:

| $n$ | $F_6(n)$ | $U(n)$ | $T(n)$ | $1/T(n)$ |
|---|---|---|---|---|
| 6 | 1 | $-6\epsilon$ | 0 | $\infty$ |
| 5 | 6 | $-4\epsilon$ | | |
| 4 | 15 | $-2\epsilon$ | | |
| 3 | 20 | 0 | $\infty$ | 0 |
| 2 | 15 | $2\epsilon$ | region of negative | |
| 1 | 6 | $4\epsilon$ | absolute temperatures | |
| 0 | 1 | $6\epsilon$ | $\partial U/\partial T > 0$ | |

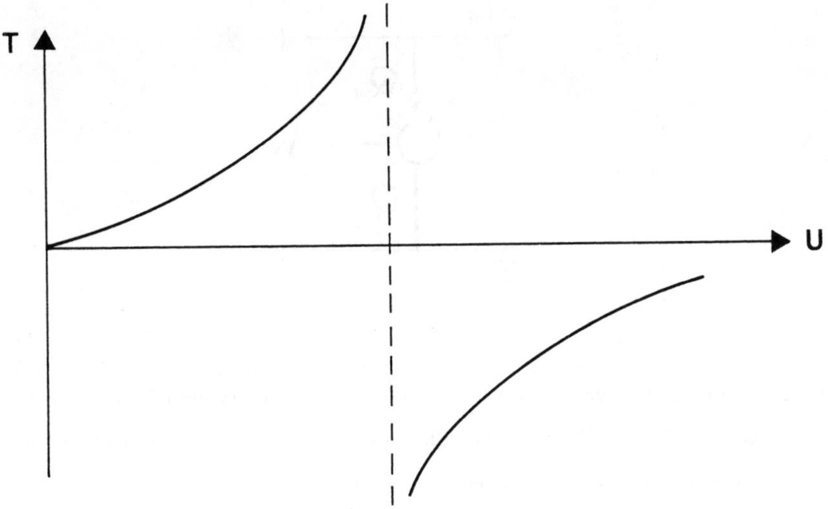


Figure 3.1: Absolute temperature graphed as function of internal energy. Negative absolute temperatures belong to higher energies than positive ones. $\partial U/\partial T > 0$ is always satisfied.

The temperature $T = 0$ belongs to the state of lowest internal energy $U(6) = -6\epsilon$, $T = \infty$ to the state $n = 3$ of the highest disorder. This state is not that of highest energy, so that states beyond $T = \infty$ exist. These states should be also characterized by a temperature under the condition that $\partial U/\partial T > 0$ is valid. As shown in Figure 3.1 only negative absolute temperature are possible. Nuclear spin systems were investigated firstly in 1951 by Purcell and Pound [11,12].

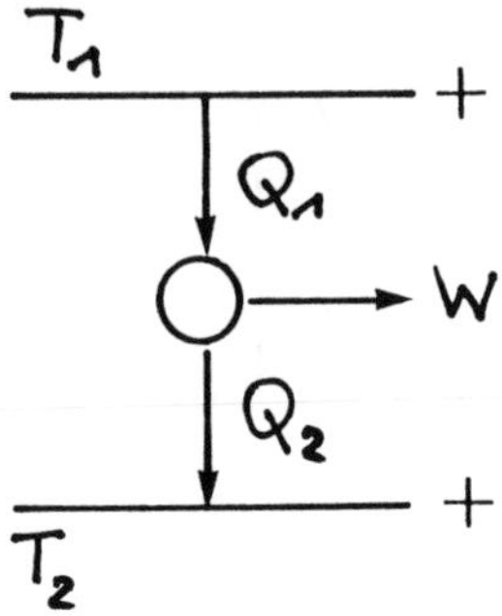


Figure 3.2: Diagram of a cyclic process between two heat reservoirs of the thermostatic temperatures $T_1$ and $T_2$, $T_1 > T_2$. The heat exchanges are $Q_1$ and $Q_2$, $W$ is the power exchange. The first law writes $Q_1 + Q_2 + W = 0$.

## 3.3 Diagram Technique

For formulating the second law including also negative absolute temperatures we use a special diagram technique [13]. Figure 3.2 shows a diagram of a system undergoing a cyclic process which during this process is contacted with two equilibrium heat reservoirs represented by horizontal lines. The heat exchanges between the considered system and the heat reservoirs are represented by perpendicular arrows. The horizontal arrow symbolizes the total work exchange of the cyclic process.

## 3.4 Axioms

Starting out with verbal formulations of the second law an analytical expression — Clausius' inequality — is proved also including negative absolute temperatures.

### 3.4.1 Positive Temperatures

Experience shows the following

*Empirem*: There are no Kelvin- and Clausius-processes, but heat conduction- and friction-processes do exist.

The diagrams belonging to this empirem are in the 2-reservoir case.

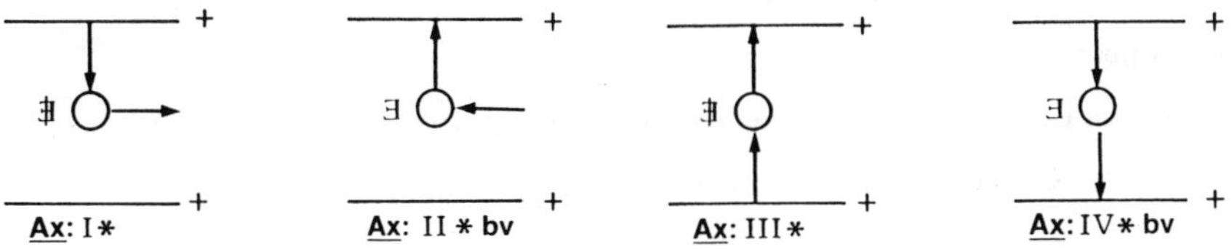


Here ∃ means this diagram is existing, i.e. such a cycle process can be observed, ∄ means that the process belonging to the diagram is not observable in nature. The +-signs denote heat reservoirs of positive absolute temperature. The diagrams are enumerated by roman digits.

### 3.4.2 Negative Temperatures

According to Figure 3.1 negative absolute temperatures belong to higher internal energies. Heat conducting processes satisfy the

*Empirem*: Also in case of including negative absolute temperatures the heat exchange in heat conducting processes is directed from systems of higher specific internal energy to those of lower specific internal energy.

The diagrams belonging to this empirem are in case of positive absolute temperatures III and IV, and in general:

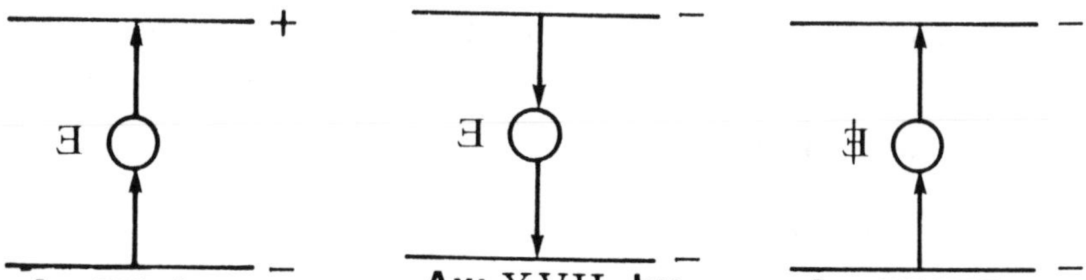


**Ax:** XI, **bv** **Ax:** XVII, **bv** **Ax:** XVIII

In spin systems of negative absolute temperatures heat conduction is caused by the spin-lattice interaction. Here the lattice acts on the spins as heat bath of positive absolute temperature (diagram XI). But another interpretation of the role of the lattice is possible:

It acts as external work storage absorbing energy from the spin system. In the case diagram XI changes into the diagram

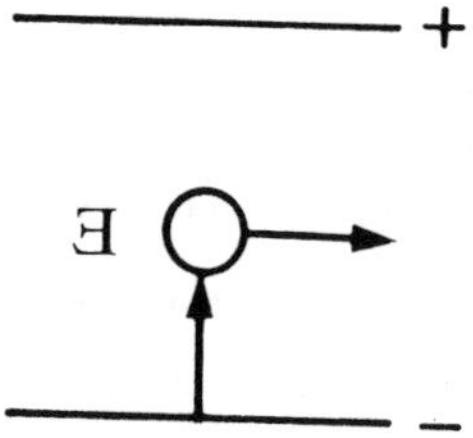


**Ax:** IX,**bv**

which is also an axiom. Therefore eight diagrams I to IV, and IX, XI, XVII, and XVIII are the axioms we start with (the somewhat strange enumeration originates of a special scheme which will be clear below).

## 3.5 Combination of Diagrams

Because heat and power exchanges are additive we can add diagrams. This means different systems can be composed to a new system because the general definition of a system allows their combination. If we combine the existent diagrams II and IV we get another existent diagram numbered by V:

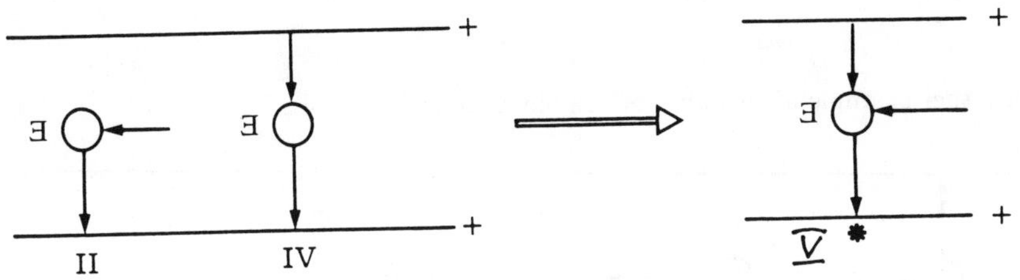


Another example is:

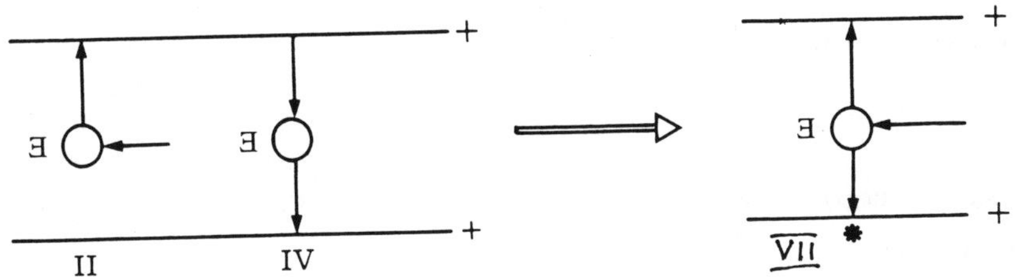


By these combinations a map

$$(\cdot\ ,\ \cdot) : \exists \times \exists \to \exists \tag{3.4}$$

is established, where $\exists$ is the set of all existing diagrams.

Another sort of combination can be made with non-existing diagrams: Starting out with the diagrams II and IV, IV exists and II is non-existing, we get by combination a non-existing diagram (VIII):

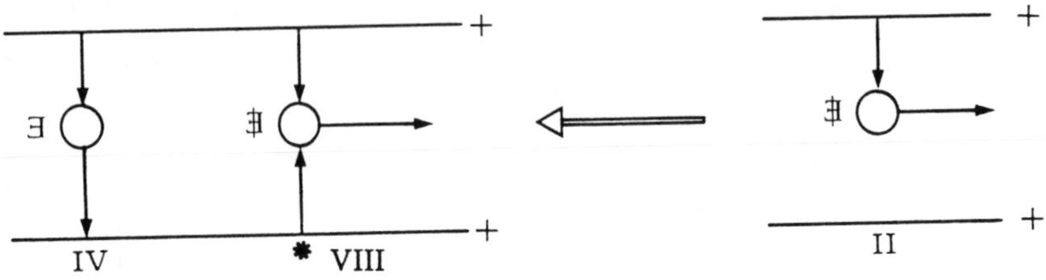


Another example of this kind of combination is:

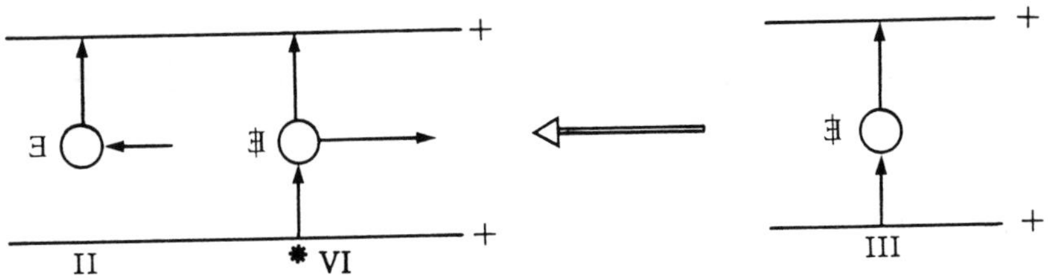


Therefore a map

$$\langle \cdot \ , \ \cdot \rangle : \exists \times \ \exists \rightarrow \exists \tag{3.5}$$

Including negative absolute temperatures and applying the maps $(\cdot,\cdot)$ and $\langle\cdot,\cdot\rangle$ we get further diagrams

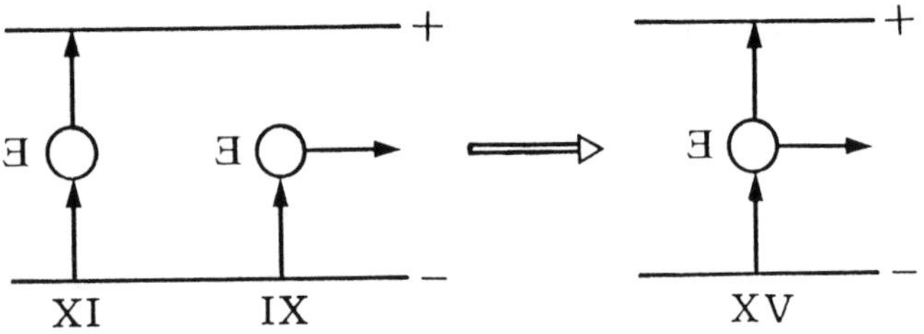

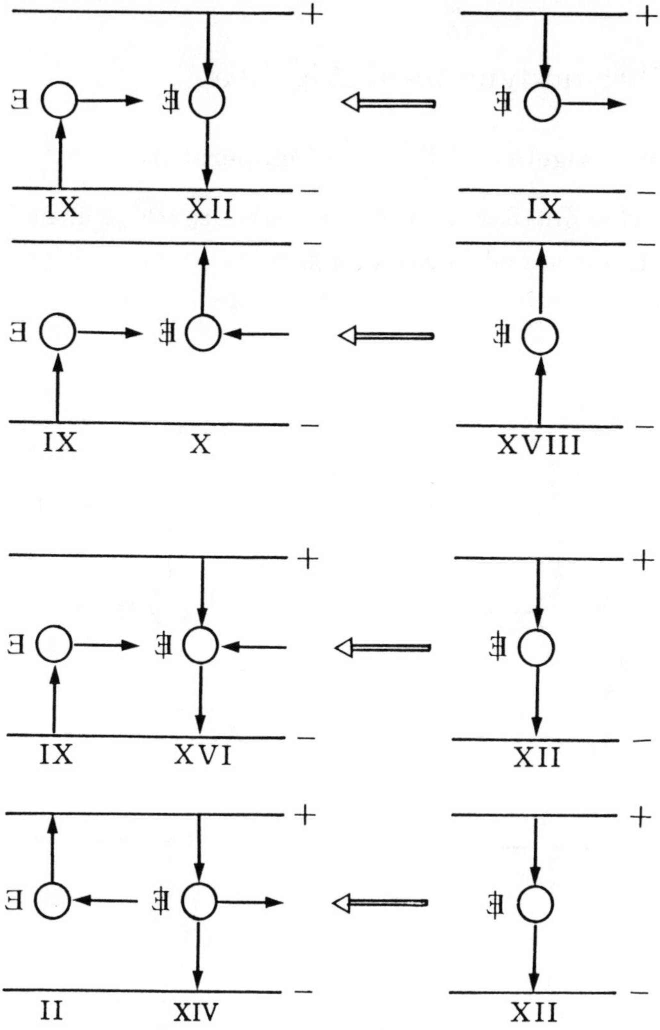
∃
∄
∄
IX
XII
IX
∃
∄
∄
IX
X
XVIII
∃
∄
∄
IX
XVI
XII
∃
∄
∄
II
XIV
XII

The set which is generated from the 8 axiom diagrams I–IV, IX, XI, XVII, and XVIII by applying $(\cdot,\cdot)$ and $\langle\cdot,\cdot\rangle$ and which is closed under these maps consists of 22 diagrams. They form an algebra which we call the *thermodynamical algebra* $\mathcal{A}$.

## 3.6 Thermodynamical Algebra

### 3.6.1 Sub-Algebra of Positive Temperatures

First of all we write down the 8 diagrams containing only positive absolute temperatures. They form a sub-algebra $\mathcal{A}^+$ of $\mathcal{A}$:

$\mathcal{A}^+$ :

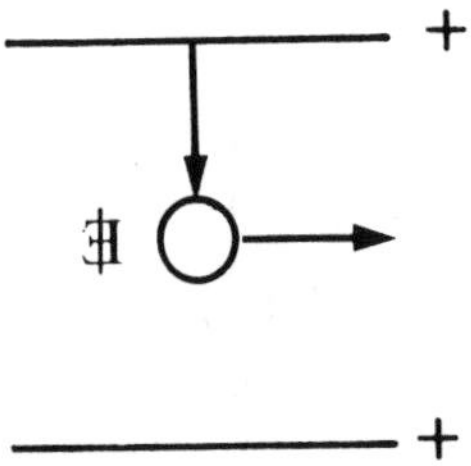

Ax: I*

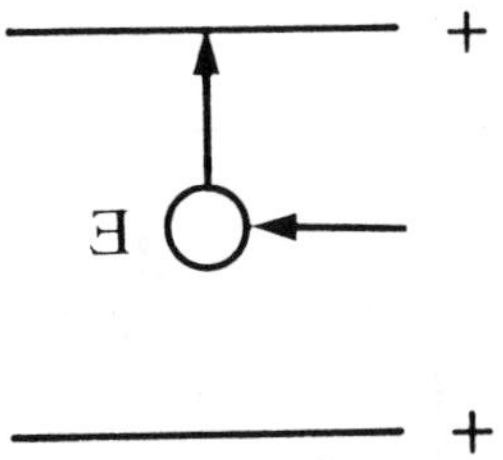

Ax: II * bv

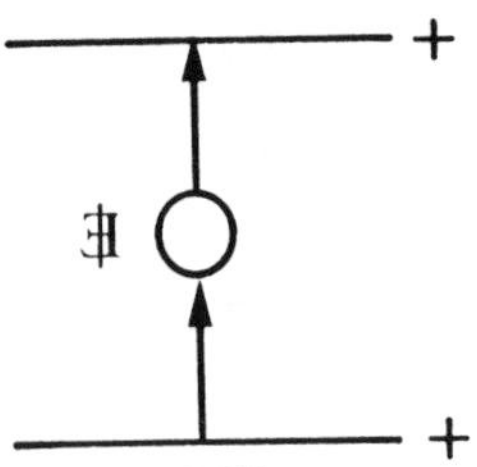

Ax: III *

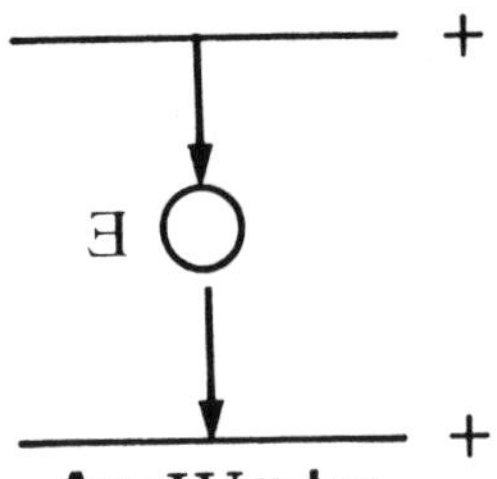

Ax: IV * bv

## 3.6.3 General case [illegible]

The general case results from [illegible] by adding all diagrams contain [illegible] alternate [illegible] of we have 8 [illegible] following [illegible] diagrammatic elements:

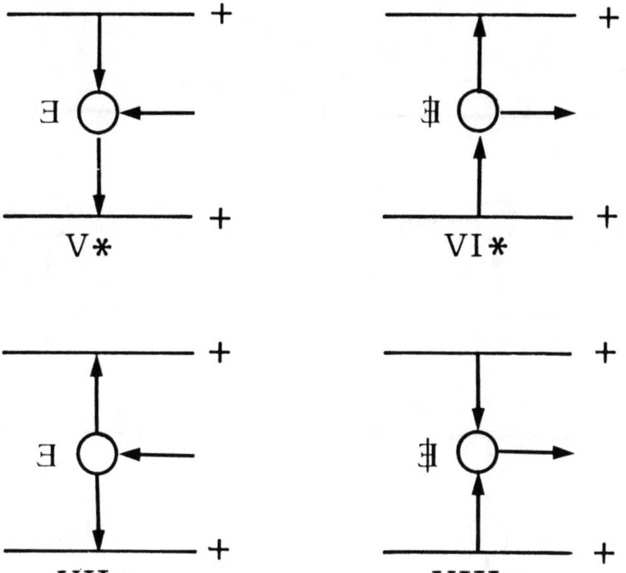

V*  VI*  VII*  VIII*

### 3.6.2 General Case

The general case results from $\mathcal{A}^+$ by adding all diagrams containing negative absolute temperatures. First we have 8 diagrams including only one reservoir of negative absolute temperature:

$\exists^+_- \cup \nexists^+_-$:

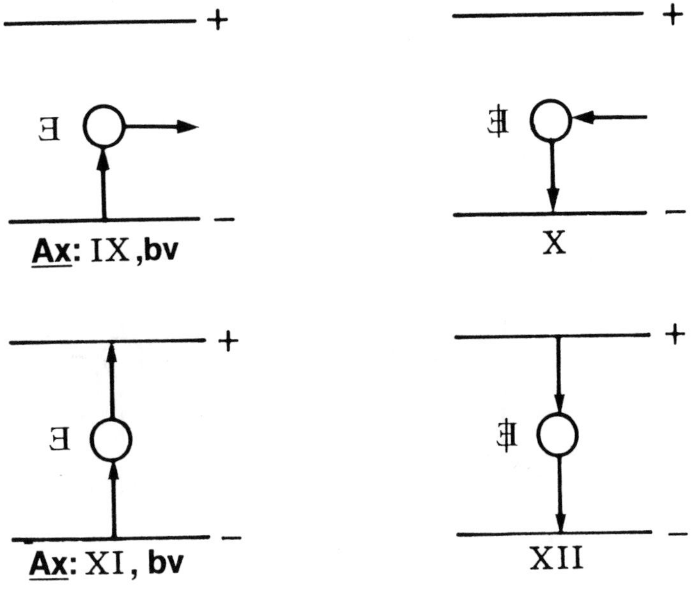

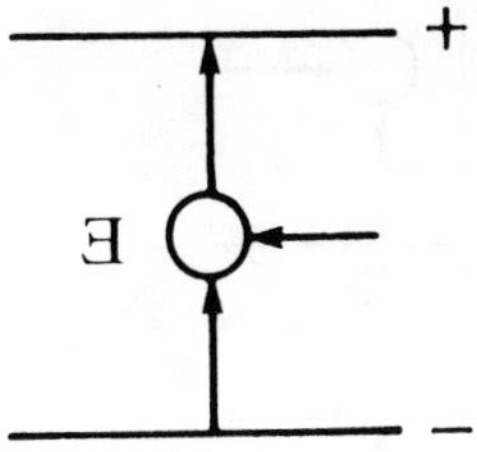


XIII

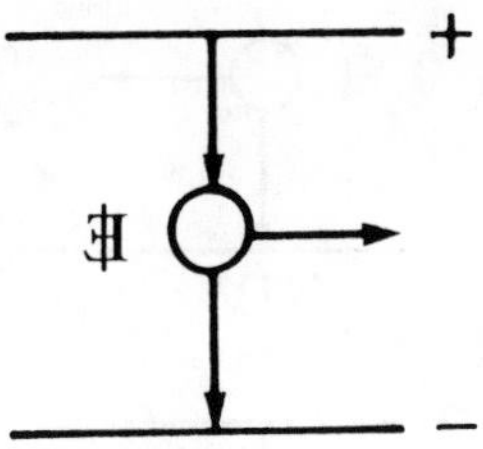


XIV

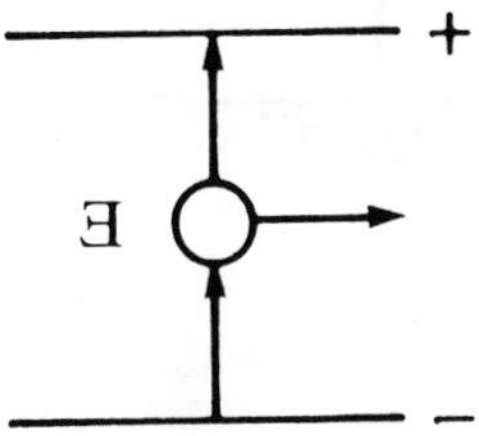


XV

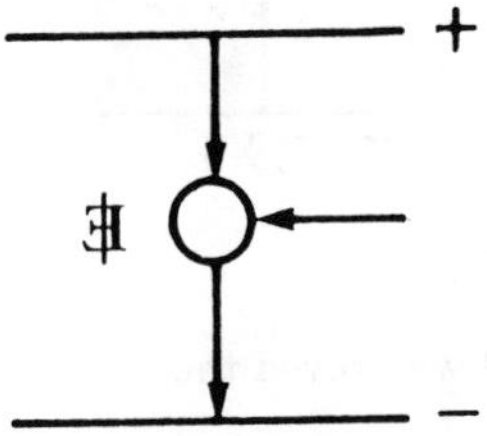


XVI

Finally there are 6 diagrams having two reservoirs of negative absolute temperatures

∃: ∪ ∄:

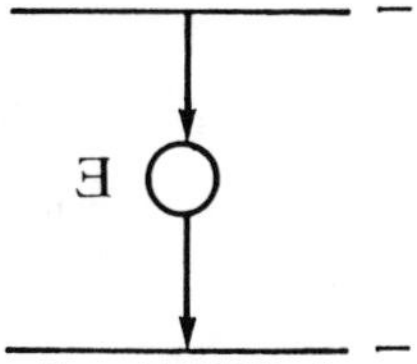


**Ax**: XVII, **bv**

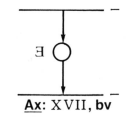

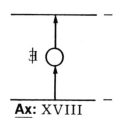

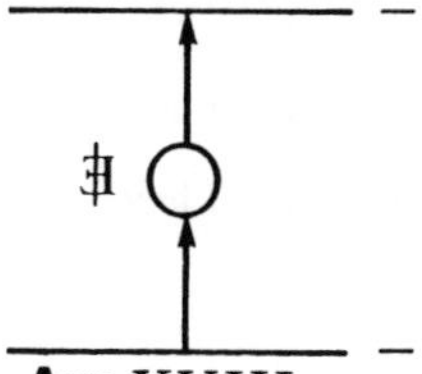


**Ax**: XVIII

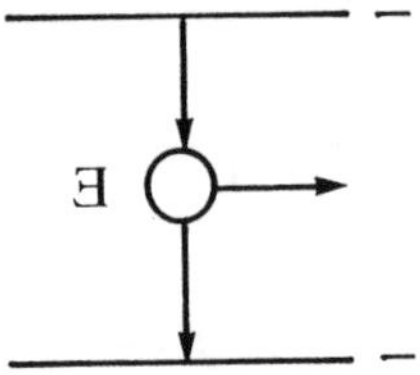


XIV

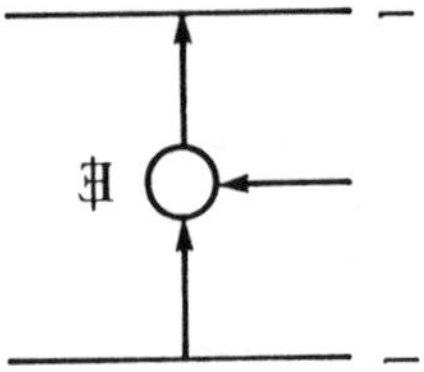


XX

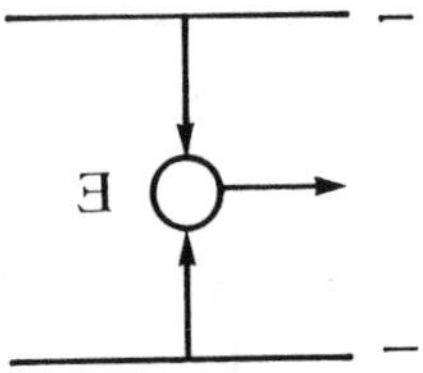


XXI

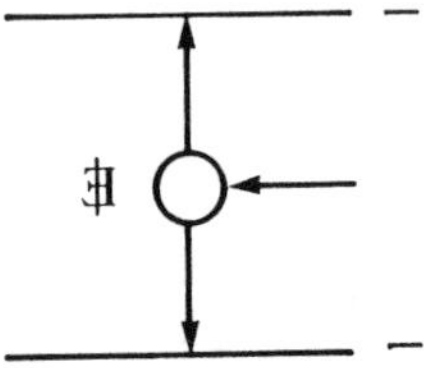


XXII

All in all we proved the

*Proposition*: The algebra $\mathcal{A}$ of the thermodynamical diagrams induced by the eight diagrams being axioms and generated by the relations $(\cdot,\cdot)$ and $\langle\cdot,\cdot\rangle$ consists of 22 diagrams. $\mathcal{A}$ is called thermodynamical algebra. The diagrams belonging to only positive heat reservoirs form a sub-algebra $\mathcal{A}^+$. $\mathcal{A}^+$ consists of 8 diagrams. All diagrams of $\mathcal{A}$ do exist or do not.

## 3.7 Complement of the Thermodynamical Algebra

Not all diagrams which are compatible with the first law are included in $\mathcal{A}$. Seven diagrams remain over. They form $\overline{\mathcal{A}}$, the complement of $\mathcal{A}$ with reference to all possible diagrams:

$\overline{\mathcal{A}}$:

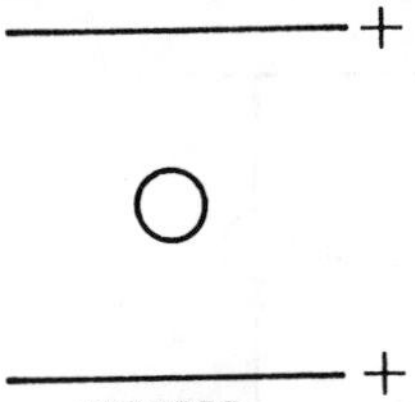


XXIII

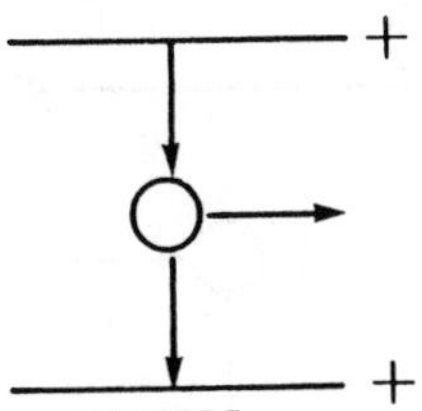


XXIV

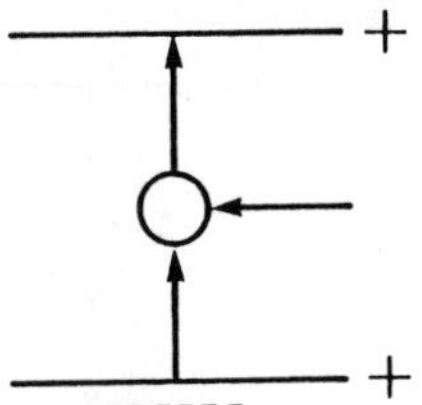


XXV

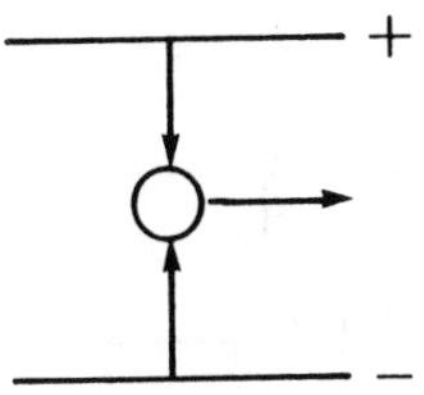


XXVI

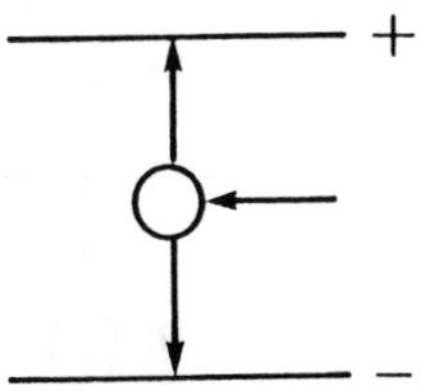


XXVII

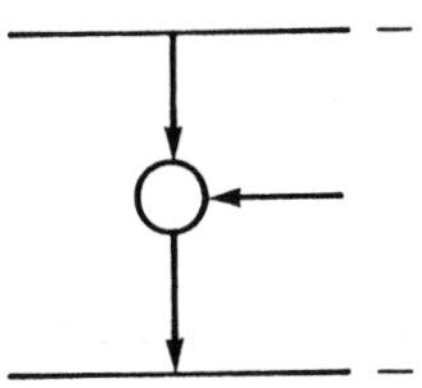


XXVIII

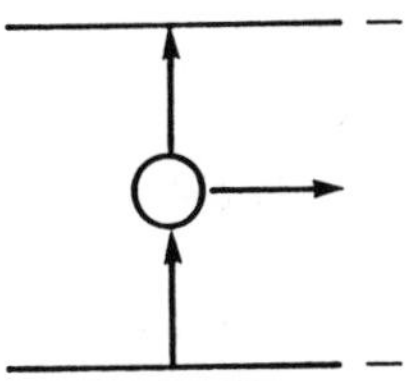


XXIX

In contrast to all diagrams of $\mathcal{A}$ all diagrams of $\overline{\mathcal{A}}$ may exist or may not. The question now is how to get conditions for deciding whether a diagram exists or does not exist. These conditions represent just the second law. So diagram XXIV symbolizes a heat-power machine, and we are looking for conditions for the existence of the diagram. For this purpose we need of course additional axioms because XXIV is not an element of $\mathcal{A}$.

### 3.7.1 Basis

All diagrams $\mathcal{A} \cup \overline{\mathcal{A}}$ form a *vector space* whose basis $\mathcal{B}$ consists of 5 diagrams of $\mathcal{A}$:

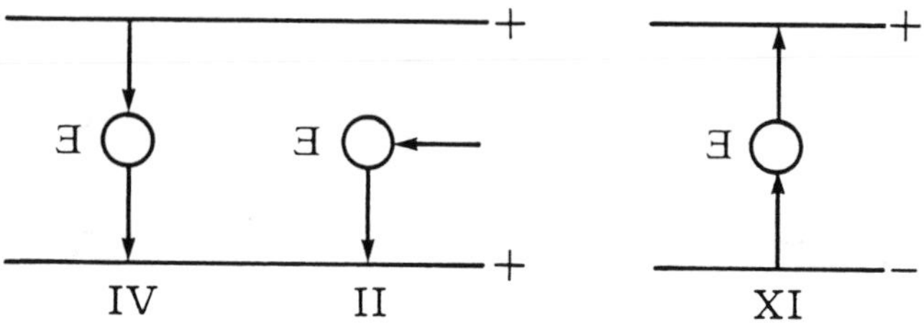


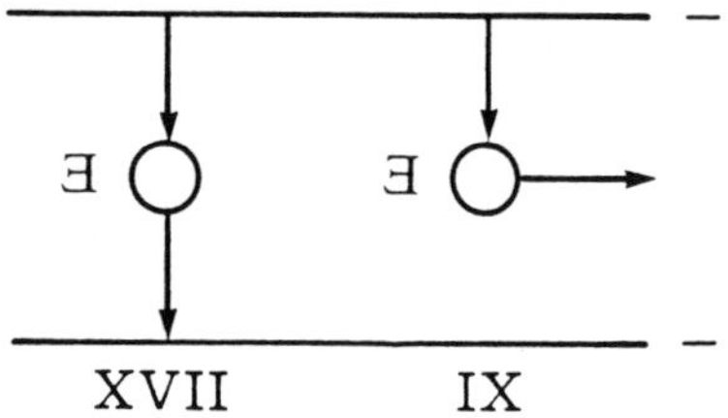


The multiplication

$$\alpha \in R \wedge b \in \mathcal{A} \cup \overline{\mathcal{A}} \rightarrow \alpha b \in \mathcal{A} \cup \overline{\mathcal{A}} \tag{3.6}$$

means that all exchanges (the arrows of the diagram) are transformed into exchanges multiplied by a real number $\alpha$. The addition

$$a, b \in \mathcal{A} \cup \overline{\mathcal{A}} \rightarrow a + b \in \mathcal{A} \cup \overline{\mathcal{A}} \tag{3.7}$$

is defined as in the maps $(\cdot,\cdot)$ and $\langle\cdot,\cdot\rangle$. In contrast to the application of these maps multiplication and addition destroy the possibility of classifying the diagrams into existing or non-existing ones. This becomes clear by the fact that multiplication and addition are defined on $\mathcal{A} \cup \overline{\mathcal{A}}$ which contains diagrams whose existence cannot be decided up to now. As easy to see all elements of $\overline{\mathcal{A}}$ have a special representation in $\mathcal{B}$:

$$a \in \overline{\mathcal{A}} \rightarrow a = b + c, b \in \exists, c \in \nexists \ . \tag{3.8}$$

We give two examples:

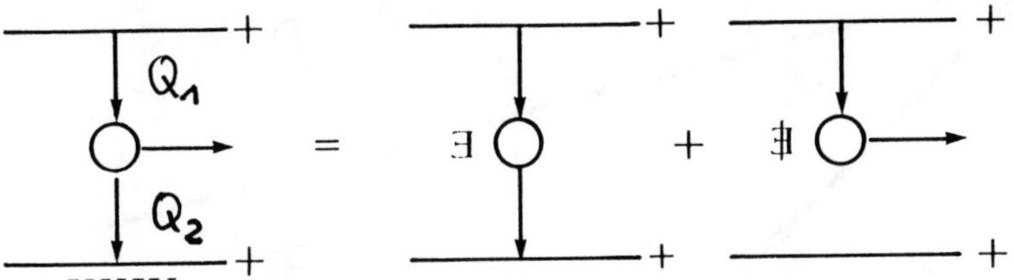


XXIV

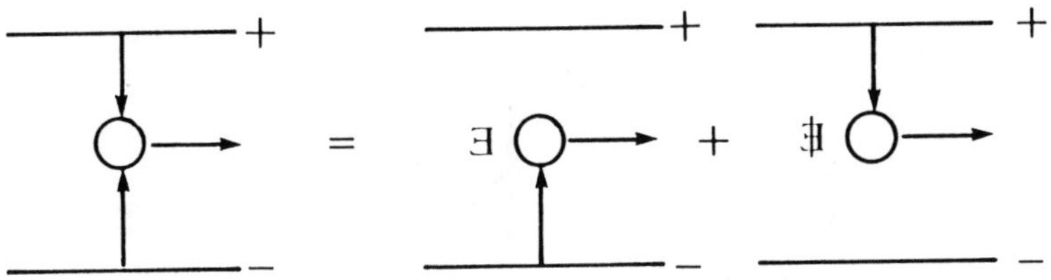


XXVI

## 3.7.2 The Size-Empirem

To proceed we now need additional empirems. The first one is the *size-empirem* which runs as follows:

*Empirem*: The existence or non-existence of a thermodynamical device (diagram) does not depend on its size.

If $\overline{\exists}$ and $\overline{\nexists}$ are the sets of existing and non-existing diagrams in $\overline{\mathcal{A}}$, the size-empirem states

$$\alpha > 0 \wedge a \in (\exists, \nexists, \overline{\exists}, \overline{\nexists}) \rightarrow \alpha a \in (\exists, \nexists, \overline{\exists}, \overline{\nexists}) \ . \tag{3.9}$$

The diagrams $a$ and $\alpha a$ are elements of the same subset $\exists$, $\nexists$, $\overline{\exists}$, or $\overline{\nexists}$ of $\mathcal{A} \cup \overline{\mathcal{A}}$. Because the "length" of the arrows cannot be seen at the diagram $a$ and $\alpha a$ are represented by the same diagram.

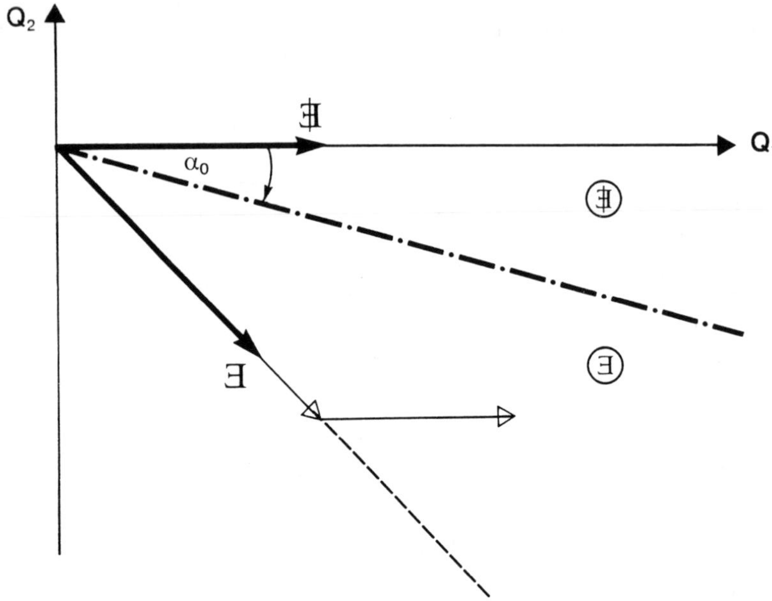


Figure 3.3: $Q_1 - Q_2$ graph of the linear combined diagram XXIV. $Q_1$ belongs to the reservoir of higher temperature. $\alpha_0$ is the angle between the $Q_1$-axis and the *ray of reversible diagrams*.

Figure 3.3 shows the $Q_1$–$Q_2$ graph of diagram XXIV which consists of a linear combination of the two basic graphs IV and I. In $(Q_1, Q_2)$ coordinate IV has the representation $(Q_{\mathrm{IV}}, -Q_{\mathrm{IV}})$, and I $(Q_I, 0)$. According to (3.8) the diagram XXIV can be represented by

$$(Q_1, Q_2) = \alpha(Q_{\mathrm{IV}}, -Q_{\mathrm{IV}}) + \beta(Q_I, 0), \qquad \alpha > 0, \beta > 0 \ . \tag{3.10}$$

If $(\alpha/\beta) >> 1$, the diagram XXIV exists, whereas in case of $(\alpha/\beta) << 1$ it does not. Therefore a separating line exists between existing and non-existing diagrams in the $Q_1$– $Q_2$-plane.

Diagrams on this separating line are called *reversible* because they do not belong to the existing nor to the non-existing diagrams. According to the size-empirem all diagrams on rays in the $Q_1$–$Q_2$-plane through (0,0) are of the same type, existing, non-existing, or reversible. Therefore all reversible diagrams are situated on rays through (0,0). Using the

*Axiom*: Combinations or reversible diagrams are reversible we get that exactly one *reversible ray* exists as in Figure 3.3 mentioned. From Figure 3.3 we read off the condition for existing diagrams:

$$tg\ \alpha \leq tg\ \alpha_o \leq 0 \ . \tag{3.11}$$

By

$$Q_1 > 0, \qquad Q_2 < 0 \tag{3.12}$$

(3.11) yields

$$Q_2 - (Q_2^o/Q_1^o)Q_1 \leq 0 \ . \tag{3.13}$$

This inequality proved for diagram XXIV is a forerunner of Clausius' inequality.

The angle $\alpha_o$ (Figure 3.3) of the reversible ray depends only on the empirical temperatures $\theta_1$ and $\theta_2$ of the process controlling heat reservoirs. This results from the fact that on the reversible ray the ratio $Q_2^o/Q_1^o$ is fixed and the exchanges $Q_1$ and $Q_2$ are not involved. Therefore we get

$$tg\ \alpha_o = Q_2^o/Q_1^o = f(\theta_2, \theta_1) \ . \tag{3.14}$$

## 3.8 Clausius' Inequality in the Case of Two Reservoirs

The last axiom states a property of combinations of reversible diagrams. We now formulate an axiom on decomposition of such diagrams:

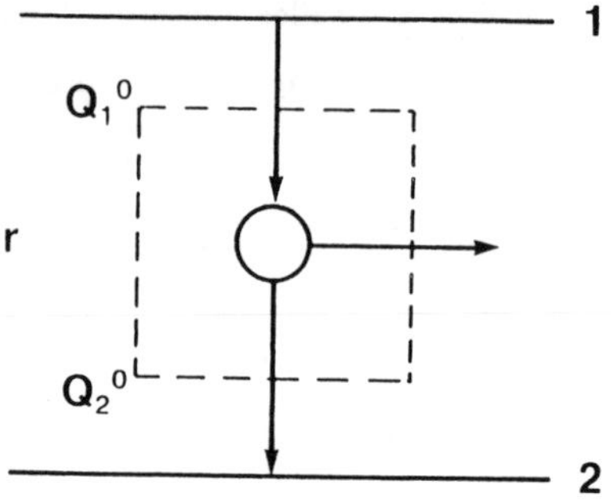


Figure 3.4: A reversible diagram between two reservoirs 1 and 2. The ratio of the heat exchanges is $Q_2^o/Q_1^o = f(\theta_2, \theta_1)$.

*Axiom*: A reversible diagram can only consist of reversible sub-diagrams.

If we consider a reversible diagram as in Figure 3.4, then the central part of the diagram may consist of other reversible diagrams as Figure 3.5 shows.

Analogical to (3.14) we get (Figure 3.5)

$$Q_3^o/Q_1^o = f(\theta_3, \theta_1)/Q_2^o - Q_3^o = f(\theta_2, \theta_3) \tag{3.15}$$

This yields

$$f(\theta_2, \theta_3)f(\theta_3, \theta_1) = -f(\theta_2, \theta_1) \tag{3.16}$$

Using a well known standard procedure [14], it follows that

$$Q_2^o/Q_1^o = f(\theta_2, \theta_1) = -T(\theta_2)/T(\theta_1) \ , \tag{3.17}$$

Here $T : R \to R$ is a monotonic map which is called the ***absolute temperature***. Because reversible diagrams do not exist $T$ has to be determined by experience. Equation (3.17) is Carnot's theorem, but nothing about the "existence" of reversible processes or of Carnot machines was presupposed.

Equations (3.13) and (3.17) yield

$$Q_2 + (T_2/T_1)Q_1 \leq 0 \ . \tag{3.18}$$

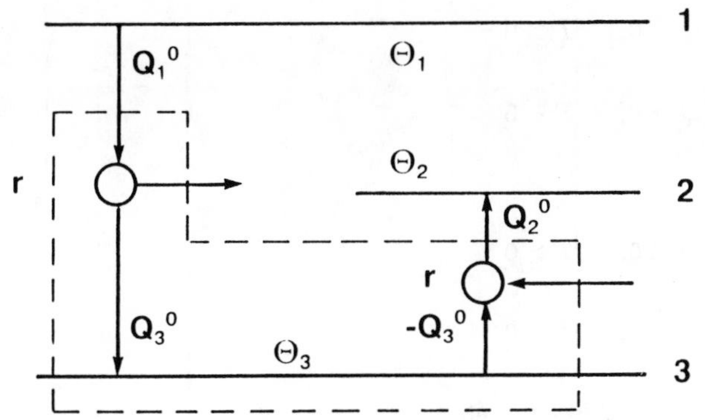


Figure 3.5: The central part of Figure 3.4 consists of two other reversible diagrams with an additional heat reservoir 3.

Because $T_2 > 0$ in XXIV we get Clausius' inequality of a 2-reservoir system:

$$\frac{Q_1}{T_1} + \frac{Q_2}{T_2} \leq 0 \ . \tag{3.19}$$

This inequality was proved starting out with diagram XXIV, but it holds in general for 2-reservoir systems:

*Problem 3.1*: Clausius's inequality (3.19) is also valid for existing diagrams XXV to XXIX.

## 3.9 The General Case

The generalizing (3.19) to arbitrary closed systems we consider Clausius' integral which using the mean value theorem we divide into four parts, each of which is characterized by the sign of the rate of heat exchange $\dot{Q}$ and that of the temperature during the process:

$$\oint \frac{\dot{Q}(t)}{T(t)} dt = \frac{Q^{++}}{T^{++}} + \frac{Q^{+-}}{T^{+-}} + \frac{Q^{-+}}{T^{-+}} + \frac{Q^{--}}{T^{--}} \tag{3.20}$$

with

$$++ : \dot{Q} \geq 0, \qquad T > 0$$

$$+- : \dot{Q} \geq 0, \qquad T < 0$$

$$(3.21)$$

$$-+ : \dot{Q} < 0, \qquad T > 0$$

$$-- : \dot{Q} < 0, \qquad T < 0$$

$$Q^{++} := \int_{++} \dot{Q} dt \geq 0 \text{ , etc.} \tag{3.22}$$

$T^{++}, T^{+-}, \ldots$ mean values on $++, +-, \ldots$

Applying the mean value theorem again we get

$$\frac{Q^{++}}{T^{++}} + \frac{-Q^{--}}{-T^{--}} = \frac{Q_1}{T_1} \ , \tag{3.23}$$

with

$$Q_1 := Q^{++} - Q^{--} \geq 0 \ , \tag{3.24}$$

$$T^{++} \ldots T_1 \ldots - T^{--} \ , \tag{3.25}$$

and

$$\frac{Q^{-+}}{T^{-+}} + \frac{-Q^{+-}}{-T^{+-}} = \frac{Q_2}{T_2} \ , \tag{3.26}$$

$$Q_2 := Q^{-+} - Q^{+-} < 0 \ , \tag{3.27}$$

$$T^{-+} \ldots T_2 \ldots - T^{+-} \ . \tag{3.28}$$

Therefore (3.20) becomes

$$\oint \frac{\dot{Q}(t)}{T(t)} dt = \frac{Q_1}{T_1} + \frac{Q_2}{T_2} \ , \tag{3.29}$$

and we proved

*Proposition:* The value of Clausius' integral including negative absolute temperatures can be represented by the diagrams XXIV or XXV, both belonging to $\overline{\mathcal{A}}$ and having only reservoirs of positive absolute temperatures.

For determining the sign of Clausius' integral we make the *assumption of contradiction*

$$\oint \frac{\dot{Q}(t)}{T(t)} dt > 0 \ . \tag{3.30}$$

Then a cyclic 2-reservoir process exists satisfying

$$\hat{Q}_1 = -Q_1, \hat{Q}_2 = -Q_2 \ , \tag{3.31}$$

$$\frac{\hat{Q}_1}{T_1} + \frac{\hat{Q}_2}{T_2} < 0 \ . \tag{3.32}$$

This process and the original one can be composed to a compound system which according to

$$\oint \frac{\dot{Q}(t)}{T(t)} dt + \frac{\hat{Q}_1}{T_1} + \frac{\hat{Q}_2}{T_2} = 0 \tag{3.33}$$

is reversible. Therefore according to the first axiom in 3.8 the original process must be reversible, too. But it was presupposed to be existent, and therefore it cannot be reversible so that the contradiction assumption (3.30) is wrong. Consequently we proved Clausius' inequality for arbitrary closed systems including also negative absolute temperatures

$$\oint \frac{\dot{Q}(t)}{T(t)} dt \leq 0 \tag{3.34}$$

Finally we put together all 29 numbered diagrams of the thermodynamical algebra $\mathcal{A}$ and its complementary set $\overline{\mathcal{A}}$ in a table below which shows the number of diagrams belonging to special subsets:

Basis ↓

| | $\mathfrak{G}$ | | | $\mathcal{A}\cup\bar{\mathcal{A}}$ | | $\mathcal{A}\cup\bar{\mathcal{A}}$ | | |
|---|---|---|---|---|---|---|---|---|
| | $\mathfrak{B}$ | | | $\mathcal{A}^+\cup\bar{\mathcal{A}}^+$ | | | | |
| $\exists^+_+$ | 2 | | 2 | 4 | | | | |
| $\exists^+_-$ | 2 | | 2 | | 4 | 11 | $\exists$ | |
| $\exists^-_-$ | 1 | | 2 | | 3 | | | $\mathcal{A}$ |
| $\nexists^+_+$ | | 2 | 2 | 4 | | | | |
| $\nexists^+_-$ | | | 4 | | 4 | 11 | $\nexists$ | |
| $\nexists^-_-$ | | 1 | 2 | | 3 | | | |
| $\bar{\mathcal{A}}^+_+$ | | | 3 | 3 | | | | |
| $\bar{\mathcal{A}}^+_-$ | | | 2 | | 2 | 7 | $\bar{\mathcal{A}}$ | |
| $\bar{\mathcal{A}}^-_-$ | | | 2 | | 2 | | | |
| | 5 | 3 | 21 | 11 | 18 | 29 | | |

Rows $\exists^+_+$ to $\nexists^-_-$: Thermodynamical Algebra

Rows $\bar{\mathcal{A}}^+_+$ to $\bar{\mathcal{A}}^-_-$: Its Complement

Columns 5, 3: Axioms

Column 21: Diagrams derived from the axioms

Column 11: T > 0

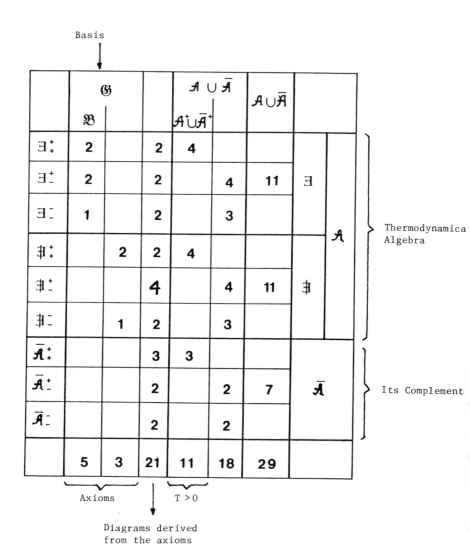

# Chapter 4

# Non-Equilibrium Contact Quantities and their Quantum-Statistical Foundation

## 4.1 Preliminary Considerations

### 4.1.1 "Dynamic" Pressure

We consider an isolated system consisting of two subsystems (Figure 4.1). Both subsystems are separated by a moveable adiabatic piston which is impervious to material. At the beginning one subsystem is in equilibrium, the other is in an arbitrary state. Introducing the mean pressure by

$$\int_{\text{piston}} P(\underline{x}, t)\dot{\underline{x}} \cdot d\underline{f} = \overline{P} \int_{\text{piston}} \dot{\underline{x}} \cdot d\underline{f} = \overline{P}\dot{V} \ , \tag{4.1}$$

we get for the initial time the inequality

$$(\overline{P} - P^{\text{eq}})\dot{V} \geq 0 \ . \tag{4.2}$$

Of course $\overline{P}$ and $\dot{V}$ are functions of time. If we contact the non-equilibrium system at time $t$ with a suitable equilibrium system of the pressure $P^{\text{eq}}(t)$, (4.2) is also valid for all times

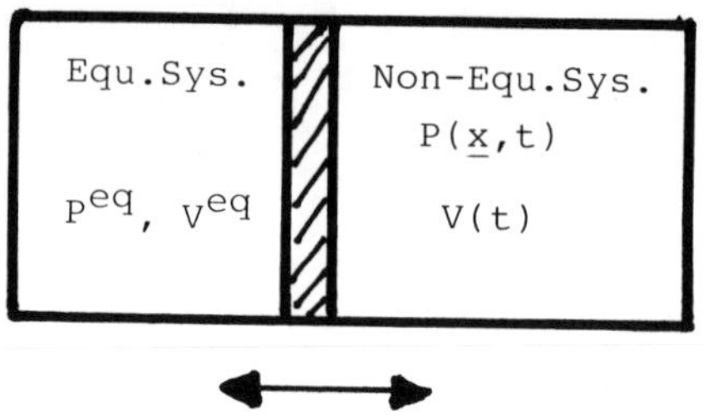


Figure 4.1: An isolated system consisting of two subsystems separated by a partition which is only permeable to work.

$$[\overline{P}(t) - P^{\mathrm{eq}}(t)]\dot{V}(t) \geq 0 \ . \tag{4.3}$$

We interpret this inequality as follows: The "dynamic" pressure $\overline{P}(t)$ of the non-equilibrium system at time $t$ can be measured by the pressure $P^{\mathrm{eq}}(t)$ of an equilibrium system being contacted with the non-equilibrium system by a partition only permeable to work. Both pressures are equal

$$\overline{P}(t) = P^{\mathrm{eq}}(t) \ , \tag{4.4}$$

if

$$\dot{V}(t) = 0 \ . \tag{4.5}$$

Therefore $\dot{V} = 0$ is the indicator for the equality of the dynamic pressure and the pressure of the equilibrium system. By this procedure the non-equilibrium pressure can be defined by the equilibrium pressure.

Generalizing pressure and volume to the generalized forces and the work variables (Section 1.1)

$$P \rightarrow \underline{A} \ , \qquad \dot{V} \rightarrow -\underline{\dot{a}} \tag{4.6}$$

we get the *defining inequality* for the generalized dynamic forces

$$[A_k^{\mathrm{eq}}(t) - A_k(t)]\dot{a}_k(t) \geq 0 \ , \quad k = 1, 2, \ldots, w \ . \tag{4.7}$$

Here $A_k^{\mathrm{eq}}$ is the equilibrium quantity for gauging, $\dot{a}_k$ the indicator, and $A_k$ is the defined non-equilibrium quantity.

### 4.1.2 Contact Temperature

We consider two subsystems of an isolated system which are separated by a partition being only permeable for heat exchange. As in Section 4.1.1 one subsystem is in equilibrium at the beginning and the other subsystem is an arbitrary state. Because of the chosen partition here the heat exchange between both subsystems is the indicator, and the defining inequality for the "dynamic" temperature $\theta$ of the non-equilibrium system writes

$$\left[\frac{1}{\theta}(t) - \frac{1}{T}(t)\right] \dot{Q}(t) \geq 0 \ . \tag{4.8}$$

$T$ is the thermostatic temperature of the contacting equilibrium system at time $t$. $\theta(t)$ is called *contact temperature* of the thermal contact between equilibrium and non-equilibrium system [15]. The contact temperature depends on individual properties of this thermal contact. In equilibrium $\theta = T$ holds independently of its special nature.

As easy to see $\theta$ and $U$ are independent of each other: Two subsystems of an isolated systems exchange only heat (Figure 4.2). One subsystem is controlled by different other subsystems which are always in equilibrium. We now choose the thermostatic temperature of the controlling reservoirs so that $\dot{Q}(t) \equiv 0$. This special choice is achieved by $T(t) = \theta(t)$. Because $\dot{Q} = 0$ we get for this partition $\dot{U} = 0$ according to the first law, but $\theta(t)$ depends on $t$ because NES is not in equilibrium. Consequently $U$ and $\theta$ are independent of each other.

We now introduce the minimal (non-equilibrium) state space [16]

$$Z = (\underline{a}, \underline{n}, U, \theta) \ . \tag{4.9}$$

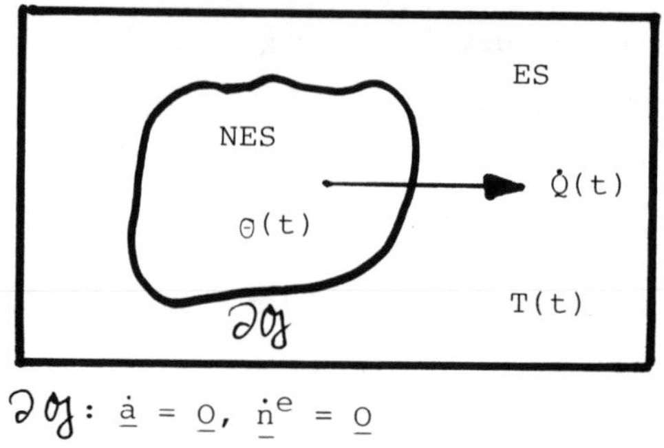


Figure 4.2: A non-equilibrium system (NES) of the contact temperature $\theta(t)$ is contacted with a set of equilibrium systems (ES) of the thermostatic temperature $T(t)$. The partition $\partial \mathcal{G}$ is permeable only for heat.

In equilibrium we get

$$Z_o = (\underline{a}, \underline{n}, U, T) \ , \quad T = C(\underline{a}, \underline{n}, U) \ , \tag{4.10}$$

where $C$ is the caloric equation of state in equilibrium. Because in equilibrium the contact temperature becomes dependent on the variables of the equilibrium subspace its dimension is

$$\dim\{Z_o\} = \dim\{Z\} - 1 \ . \tag{4.11}$$

Two special projections can be introduced onto minimal state space: The $U$-projection $P^U$ defined as

$$P^U(\underline{a}, \underline{n}, U, \theta) = (\underline{a}, \underline{n}, U) \ , \tag{4.12}$$

and the $\theta$-projection $P^\theta$

$$P^\theta(\underline{a}, \underline{n}, U, \theta) = (\underline{a}, \underline{n}, U^\theta) \ , \tag{4.13}$$

where $U^\theta$ is defined by the caloric equation of state

$$\theta = C(\underline{a}, \underline{n}, U^\theta) \ . \tag{4.14}$$

### 4.1.3 Chemical Potentials

We now consider two subsystems of an isolated system, one in equilibrium, the other in an arbitrary state. The thermostatic temperature is equal to the contact temperature, and also the generalized forces have the same value:

$$\theta(t) = T(t) \to \dot{Q}(t) = 0 \ , \tag{4.15}$$

$$\underline{A}(t) = \underline{A}^{\mathrm{eq}}(t) \to \underline{\dot{a}}(t) = \underline{0} \ . \tag{4.16}$$

Here the indicators according to Section 4.1.1 are the external rates of the mole numbers. The defining inequality of the "dynamical" chemical potential is

$$[\mu_k^{\mathrm{eq}}(t) - \mu_k(t)]\dot{n}_k^e \geq 0$$

$$k = 1, 2, \ldots, K \ , \tag{4.17}$$

with $K$ as the number of components. A formalization and a strict treatment of the contact quantities here introduced is possible [17,18].

## 4.2 Hamiltonian of a Compound System

We consider a non-equilibrium system being in contact with an equilibrium system. The contact is general so that all exchanges may occur

$$\dot{Q} \neq 0 \ , \quad \dot{W} \neq 0 \ , \quad \underline{\dot{n}}^e \neq \underline{0} \ . \tag{4.18}$$

We denote the equilibrium subsystem by 1, the non-equilibrium subsystem by 2, and their interaction by 12. The Hamiltonian of whole the system then writes

$$\mathcal{H} = \mathcal{H}^1 + \mathcal{H}^2 + \mathcal{H}^{12} \ . \tag{4.19}$$

$\mathcal{H}^1$ and $\mathcal{H}^2$ depend only on the work variables of the subsystems

$$\mathcal{H}^1 = \mathcal{H}^1(\underline{a}^1) \ , \qquad \mathcal{H}^2 = \mathcal{H}^2(\underline{a}^2) \ . \tag{4.20}$$

The total system is isolated. Consequently we get

$$\partial \mathcal{H} / \partial t = 0 \ . \tag{4.21}$$

Besides the Hamiltonians we have the particle number operators

$$\underline{\mathcal{N}} = \underline{\mathcal{N}}^1 + \underline{\mathcal{N}}^2 \ , \tag{4.22}$$

and the following commutation rules based on mutual measurability of the quantities belonging to the corresponding operators

$$[\mathcal{H}^1, \mathcal{H}^2] = 0 \ , \qquad [\underline{\mathcal{N}}^1, \underline{\mathcal{N}}^2] = \underline{\underline{0}} \ , \tag{4.23}$$

$$[\underline{\mathcal{N}}^j, \mathcal{H}^k] = \underline{0} \ , \qquad j, k = 1, 2 \ , \tag{4.24}$$

$$[\mathcal{H}, \underline{\mathcal{N}}] = \underline{0} \ . \tag{4.25}$$

## 4.3 Set of Relevant Observables

As an analogue to the state space we introduce according to [19] a set

$$\mathcal{B} = (\mathcal{G}^1, \mathcal{G}^2, \ldots, \mathcal{G}^m) \tag{4.26}$$

of $m$ self-adjoint operators

$$\mathcal{G}^k = \mathcal{G}^{k+} \ , \qquad k = 1, \ldots, m \ , \tag{4.27}$$

which are the observables of the considered system. As in the definition of the state space no other observables than those in the set are taken as relevant. We call this set $\mathcal{B}$ *Beobachtungsebene*.

The beobachtungsebene represents the state variables whereas the state itself is represented by a density operator $\rho$ belonging to the microscopic description of the system. With regard to the beobachtungsebene the microscopic description is too fine, because $\rho$ contains more information about the considered system than that we can get due to the limited beobachtungsebene. Therefore $\rho$ is to be replaced by another density operator $R$

$$\rho \rightarrow R \tag{4.28}$$

belonging to $\mathcal{B}$ [20,21]. To get this so-called ***generalized canonical density operator*** a procedure maximalising entropy is applied [4,20,21] starting out with the following presuppositions

$$\text{i)}\ \langle \mathcal{G}^k \rangle = Tr(\rho \mathcal{G}^k) = Tr(R\mathcal{G}^k)\ , \qquad \text{for all } k\ , \tag{4.29}$$

$$\text{ii)}\ Tr\rho = 1 = TrR \tag{4.30}$$

iii) *Entropy* :

$$S_\rho = -kTr(\rho \ln \rho) \tag{4.31}$$

$$S_\rho \leq S_R \rightarrow \max\ .$$

Here the brackets $\langle \cdots \rangle$ denote the quantummechanical expectation value.

*Problem 4.1* : There exists exactly one operator $R$ which satisfies i) to iii):

$$R = Z^{-1} \exp(-\sum_{k=1}^{m} \lambda_k \mathcal{G}^k) \tag{4.32}$$

$$Z := Tr \exp(-\underline{\lambda} \cdot \underline{\mathcal{G}})\ , \qquad \underline{\lambda} = \underline{\lambda}(\langle \mathcal{G} \rangle) \tag{4.33}$$

## 4.4 Beobachtungsebene of a Compound System

As above we consider a compound system consisting of an equilibrium system (1) and a non-equilibrium system. As Beobachtungsebene we choose

$$\mathcal{B} = (\mathcal{H}^1, \mathcal{H} - \mathcal{H}^1, \underline{\mathcal{N}}^1, \underline{\mathcal{N}}^2)\ . \tag{4.34}$$

According to (4.32) and (4.26) we get the generalized canonical density operator

$$R = Z^{-1} \exp\{-\beta^1 \mathcal{H}^1 - \beta(\mathcal{H} - \mathcal{H}^1) - \underline{\alpha}^1 \cdot \underline{\mathcal{N}}^1 - \underline{\alpha} \cdot \underline{\mathcal{N}}^2\} \tag{4.35}$$

with

$$Z := Tr \exp A\ , \tag{4.36}$$

by use of the abbreviation

$$A := \{\ldots\} = -\beta \mathcal{H} - \underline{\alpha} \cdot \underline{\mathcal{N}} - (\beta^1 - \beta)\mathcal{H}^1 - (\underline{\alpha}^1 - \underline{\alpha}) \cdot \underline{\mathcal{N}}^1\ . \tag{4.37}$$

## 4.5 Thermal Contact between Open Systems

The energies of the subsystems of the considered compound system are:

$$\dot{E}^1 := \frac{d}{dt}\langle \mathcal{H}^1 \rangle = \frac{i}{\hbar}\langle [\mathcal{H}, \mathcal{H}^1] \rangle + \langle \frac{\partial}{\partial t}\mathcal{H}^1 \rangle \tag{4.38}$$

$$\dot{E}^2 := \frac{d}{dt}\langle \mathcal{H} - \mathcal{H}^1 \rangle = -\frac{d}{dt}\langle \mathcal{H}^1 \rangle = -\dot{E}^1 \tag{4.39}$$

Because of the isolation of the compound system the energy $E^1 + E^2$ is a conserved quantity as well as the particle number is:

$$\dot{\underline{N}}^1 := \frac{d}{dt}\langle \underline{\mathcal{N}}^1 \rangle = \frac{i}{\hbar}\langle [\mathcal{H}, \underline{\mathcal{N}}^1] \rangle = -\dot{\underline{N}}^2 \tag{4.40}$$

*Problem 4.2* : It holds

$$\langle [\mathcal{H}, \mathcal{H}^1] \rangle = 0 \ , \qquad \langle [\mathcal{H}, \underline{\mathcal{N}}^1] \rangle = 0 \ , \tag{4.41}$$

if the expectation value $\langle \ldots \rangle$ is carried out by using $R$.

From (4.39) to (4.41) follows

$$\dot{E}^1 = 0 \ , \qquad \dot{\underline{N}}^1 = 0 \ . \tag{4.42}$$

Because the partition between the subsystems was presupposed as permeable for work, heat, and material the chosen Beobachtungsebene (4.34) cannot be appropriate for all times. This elucidates from the fact the equilibrium system (1) is not in equilibrium anymore.

To proceed we need time-smoothed rates of heat- and mass-exchange:

$$^1E_{t'}^{t'+\epsilon} := \frac{i}{\hbar}\int_{t'}^{t'+\epsilon} \langle [\mathcal{H}, \mathcal{H}^1] \rangle \, dt \ . \tag{4.43}$$

The integrand yields

$$\langle [\mathcal{H}, \mathcal{H}^1] \rangle = Tr(R(t)[\mathcal{H}(t), \mathcal{H}^1(t)]) \ . \tag{4.44}$$

Using the mean value theorem in (4.43) we get

$$^{1}E_{t'}^{t'+\epsilon} = \frac{i}{\hbar}\epsilon Tr\{R(t'+\xi)[\mathcal{H}(t'+\xi), \mathcal{H}^1(t'+\xi)]\} \ , \tag{4.45}$$

$$t' < t' + \xi < t' + \epsilon \ .$$

If $\epsilon$ is small enough

$$R(t'+\xi) = R(t') + \xi\dot{R}(t') \ , \tag{4.46}$$

where $R(t')$ is the relevant part of $\rho(t')$ which is generated by a projection [22]:

$$R(t') = P\rho(t') \ , \qquad PP = P \ , \tag{4.47}$$

$$\dot{R}(t') = P\dot{\rho}(t') \ . \tag{4.48}$$

Therefore (4.46) becomes

$$R(t'+\xi) = R(t') + \xi P\dot{\rho}(t') \ . \tag{4.49}$$

The commutator in (4.45) yields by use of (4.21):

$$[\ldots,\ldots] = [\mathcal{H}(t'), \mathcal{H}^1(t') + \xi\dot{\mathcal{H}}^1(t')] =$$

$$= [\mathcal{H}(t'), \mathcal{H}^1(t')] + \xi[\mathcal{H}(t'), \dot{\mathcal{H}}^1(t')] \ . \tag{4.50}$$

*Problem 4.3* : The time-smoothed time derivative of $E^1$ is:

$$\check{E}^1(t') := \frac{1}{\epsilon}{}^1E_{t'}^{t'+\epsilon} = \xi\frac{i}{\hbar}Tr\{R(t')[\mathcal{H}(t'), \dot{\mathcal{H}}^1(t')] +$$

$$+\frac{i}{\hbar}P[R(t'), \mathcal{H}(t')][\mathcal{H}(t'), \mathcal{H}^1(t')]\} \ . \tag{4.51}$$

Because of $\dot{\underline{N}}^1 = \underline{0}$ we get the analogous expression of the time-smoothed mass-exchange:

$$\check{\underline{m}}^1(t') = -\xi\frac{1}{\hbar^2}Tr\{P[R(t'), \mathcal{H}(t')][\mathcal{H}(t'), \underline{N}^1(t')]\} \ . \tag{4.52}$$

if $\check{E}^1(t')$ is split into work- and heat-exchange

$$\dot{\mathcal{H}}^1 = (\partial\mathcal{H}^1/\partial\underline{a}^1)\cdot\dot{\underline{a}}^1 \ , \tag{4.53}$$

we get the time-smoothed rates of power-, heat-, and mass-exchange of the equilibrium system (1)

$$\check{L}^1 = \xi \frac{i}{\hbar} Tr \left\{ R \left[ \mathcal{H}, \frac{\partial \mathcal{H}^1}{\partial \underline{a}^1} \right] \cdot \underline{\dot{a}}^1 \right\} \qquad \text{power} \tag{4.54}$$

$$\check{Q}^1 = [-\xi \frac{1}{\hbar^2} Tr\{P[R, \mathcal{H}][\mathcal{H}, \mathcal{H}^1]\} \qquad \text{heat} \tag{4.55}$$

$$\underline{\check{m}}^1 = -\xi \frac{1}{\hbar^2} Tr\{P[R, \mathcal{H}][\mathcal{H}, \underline{\mathcal{N}}^1]\} \qquad \text{mass} \ . \tag{4.56}$$

From (4.55) and (4.56) we get the condition for processes being adiabatic and for partitions being inpermeable to material:

$$\text{adiabatic:} \quad [\mathcal{H}^{12}, \mathcal{H}^1] = 0 \ , \tag{4.57}$$

$$\text{impermeable:} \quad [\mathcal{H}^{12}, \underline{\mathcal{N}}^1] = \underline{0} \ . \tag{4.58}$$

*Problem 4.4* : Using only quantummechanical argumentation we get the inequality:

$$\check{Q}^1(\beta^1 - \beta) + \underline{\check{m}}^1 \cdot (\underline{\alpha}^1 - \underline{\alpha}) \geq 0 \ . \tag{4.59}$$

We now will discuss this inequality is the defining inequality of the contact temperature and of the chemical potential of the non-equilibrium subsystem.

## 4.6 Special Case: Equilibrium

Each subsystem of the compound system is in equilibrium and both are in thermal equilibrium with each other:

$$\beta^1 = \beta^2 = \beta \ , \qquad \underline{\alpha}^1 = \underline{\alpha}^2 = \underline{\alpha} \ . \tag{4.60}$$

Then the generalized canonical density operator (4.35) becomes the grand-canonical operator

$$R^{\text{eq}} = Z^{-1} \exp(\beta \mathcal{H} + \underline{\alpha} \cdot \underline{\mathcal{N}}) \ . \tag{4.61}$$

Because of being in equilibrium $\beta$ is the reciprocal thermostatic temperature and $\underline{\alpha}$ the negative chemical potentials:

$$\beta = 1/kT \ , \qquad \underline{\alpha} = -\underline{\mu}/kT \ . \tag{4.62}$$

## 4.7 Contact Quantities

We denoted the equilibrium subsystems of the considered compound system by (1), and the non-equilibrium subsystem has no special index. Then the following equations are valid:

$$\check{Q}^1 = -\check{Q} \qquad \beta^1 = 1/kT \qquad \underline{\alpha}^1 = -\underline{\mu}^{eq}/kT \tag{4.63}$$

$$\check{\underline{m}}^1 = -\check{\underline{m}} \qquad \beta = 1/k\theta \qquad \underline{\alpha} = -\underline{\mu}/k\theta \tag{4.64}$$

$$\check{Q}\left(\frac{1}{\theta} - \frac{1}{T}\right) + \check{\underline{m}} \cdot \left(\frac{\underline{\mu}^{\text{eq}}}{T} - \frac{\underline{\mu}}{\theta}\right) \geq 0 \ . \tag{4.65}$$

Here $\theta$ is the contact temperature and $\underline{\mu}$ the dynamical chemical potentials of the non-equilibrium system. The indicators are $\check{Q}$ and $\check{\underline{m}}$ as discussed in section 4.1.

## 4.8 "Dynamical" Generalized Forces

The power on the equilibrium subsystem is according to (4.38) and (4.54)

$$L^1 = \langle \partial \mathcal{H}^1/\partial t \rangle + \check{L}^1 \ , \tag{4.66}$$

and that on the non-equilibrium subsystem

$$L := \langle \partial(\mathcal{H} - \mathcal{H}^1)/\partial t \rangle \ . \tag{4.67}$$

Because $\xi$ in $\check{L}^1$ is sufficiently small (contact duration) we will neglect it. (4.66) and (4.67) become

$$L^1 = Tr\{R(\partial \mathcal{H}^1/\partial \underline{a}^1) \cdot \dot{\underline{a}}^1\} \ , \tag{4.68}$$

$$L = Tr\{R[\partial(\mathcal{H} - \mathcal{H}^1)/\partial \underline{a}^2] \cdot \dot{\underline{a}}^2\} \ . \tag{4.69}$$

The first law (2.11) is written in time rates

$$\dot{E} = \dot{Q} + L = \dot{Q} + \dot{W} + \dot{E}_{\text{kin}} \tag{4.70}$$

Because the compound system is isolated $\dot{W} = 0$ follows and the power is the time rate of the kinetic energy

$$L^1 + L = \dot{E}_{\text{kin}} = \frac{d}{dt}\phi(\kappa) \ , \tag{4.71}$$

where $\phi$ is a function of a quadratic form of the time rates of the work variables:

$$E_{\text{kin}} = \phi(\kappa) \geq 0 \ , \qquad d\phi/d\kappa \geq 0 \ , \tag{4.72}$$

$$\kappa := \underline{\dot{a}}^1 \cdot \underline{\underline{U}} \cdot \underline{\dot{a}}^1 + \underline{\dot{a}}^2 \cdot \underline{\underline{V}} \cdot \underline{\dot{a}}^2 \geq 0 \ . \tag{4.73}$$

The tensors $\underline{\underline{U}}$ and $\underline{\underline{V}}$ are constant, symmetric, and positive definite.

Because we consider an isolated compound system the work variables of the subsystems depend on each other:

$$\underline{a}^2 = \underline{R}(\underline{a}^1) \tag{4.74}$$

$$\underline{\dot{a}}^2 = \underline{\underline{S}}(\underline{a}^1) \cdot \underline{\dot{a}}^1 \tag{4.75}$$

(4.71) becomes:

$$\frac{d\phi}{d\kappa}\dot{\kappa} = \left\{ \langle \frac{\partial \mathcal{H}^1}{\partial \underline{a}^1} \rangle + \langle \frac{\partial (\mathcal{H}^2 + \mathcal{H}^{12})}{\partial \underline{a}^2} \rangle \cdot \underline{\underline{S}} \right\} \cdot \underline{\dot{a}}^1 \dot{\kappa} \ . \tag{4.76}$$

Introducing the abbreviations

$$\underline{A}^{\text{eq}} := \langle \frac{\partial \mathcal{H}^1}{\partial \underline{a}^1} \rangle \cdot \underline{\underline{S}}^{-1} \ , \tag{4.77}$$

$$\underline{A} := -\langle \frac{\partial (\mathcal{H}^2 + \mathcal{H}^{12})}{\partial \underline{a}^2} \rangle \ , \tag{4.78}$$

$$\underline{\dot{a}} \equiv \underline{\dot{a}}^2 \ , \tag{4.79}$$

we get from (4.76) the defining inequality

$$(\underline{A}^{\text{eq}} - \underline{A}) \cdot \dot{\kappa} \underline{\dot{a}} \geq 0 \ . \tag{4.80}$$

*Problem 4.5* :

$$\underline{\dot{a}} = \underline{0} \rightarrow \dot{\kappa} = 0 \ . \tag{4.81}$$

The equations (4.64) and (4.78) demonstrate that the phenomenologically defined contact quantities, as contact temperature $\theta$, dynamic chemical potentials $\underline{\mu}$, and generalized forces $\underline{A}$ are also defined quantum-statistical and their defining inequalities can be derived by usual quantum-statistically methods.

# Chapter 5

# Existence of Non-Negative Entropy Production

## 5.1 Preliminary Considerations

In Section 3 we dealt with the second law. Starting out with verbal formulated axioms about the existence or non-existence of certain cyclic processes we got Clausius' inequality including also negative absolute temperatures. But there are a lot of other formulations of the second law [?] which we can divide into two classes: Formulations which are global in time like Clausius' inequality (time integrals) and those which are local in time like the formulation: Entropy production is always non-negative. Of course both the formulations are not equivalent because from formulations local in time those being global in time can be derived but not vice versa.

Here we will mention five versions of the second law:

I. Clausius' inequality for open systems:

$$\oint \left\{ \frac{\dot{Q}(t)}{T(t)} - \underline{s}(t) \cdot \dot{\underline{n}}^e(t) \right\} dt \leq 0 \ . \tag{5.1}$$

Here $T(t)$ is the thermostatic temperature of the equilibrium vicinity and $\underline{s}(t)$ its molar entropies.

II. Entropy does not decrease in isolated systems.

By release of constraints a process

$$\mathcal{F} : A(\text{eq}) \overset{\text{isol.}}{\rightarrow} B(\text{eq}) \tag{5.2}$$

undergoes starting out from a state of equilibrium $A(\text{eq})$ to another state of equilibrium $B(\text{eq})$. Then we have

$$S_B^{\text{eq}} \geq S_A^{\text{eq}} \ . \tag{5.3}$$

Another formulation of II may be starting out from a state of non-equilibrium $C$:

$$\text{III.} \quad \mathcal{F} : C \overset{\text{isol.}}{\rightarrow} B(\text{eq}) \tag{5.4}$$

$$S_B^{\text{eq}} \geq S_C \tag{5.5}$$

Here the question arises what the definition of the non-equilibrium entropy $S_C$ is.

Another formulation of II gives a statement about the time rate of entropy

$$\text{IV.} \quad \mathcal{F} \text{ isolated } : \dot{S}(t) \geq 0 \ , \tag{5.6}$$

in which also a non-equilibrium entropy appears. If we use a field formulation for describing the system, we can formulate the second law:

V. There exists a dissipation inequality which is local in position and time:

$$\frac{\partial}{\partial t} \sum_{\alpha} \rho_\alpha \hat{s}_\alpha + \nabla \cdot \left( \frac{\underline{q}}{\theta} + \underline{k} + \sum_{\alpha} \rho_\alpha \hat{s}_\alpha \underline{V}_\alpha \right) \geq 0 \ . \tag{5.7}$$

Here we have the following fields at $(\underline{x}, t)$ of the component $\alpha$:

mass density: $\rho_\alpha$

velocity: $\underline{V}_\alpha$

specific entropy: $\hat{s}_\alpha$

heat flux density: $\underline{q}$

entropy flux density: $\underline{\phi}$

with the definition:

$$\underline{k} := \underline{\phi} - \underline{q}/\theta \ , \tag{5.8}$$

where $\theta$ is a temperature whose definition is at first as undetermined as the definition of a non-equilibrium entropy.

It is obvious that I to V are *not equivalent* to each other. Here we deal with the following question:

If we start out with the basic assumptions

i) Thermostatics is presupposed to be known,

ii) Clausius' inequality holds,

can we answer the question:

Do there exist dissipation inequalities that are local in time and position?

## 5.2 State Spaces, Processes and Projections

We consider a discrete system $\mathcal{G}$ in its equilibrium surroundings $\mathcal{G}^*$ from which it is separated by the surface $\partial\mathcal{G}$ [23]. We introduce a small state space (1.11)

$$Z := (\underline{a}, \underline{n}, U, \theta, \ldots; T^*, \underline{A}^*, \underline{\mu}^*) \tag{5.9}$$

Here $\underline{a}$ are the work variables, $\underline{n}$ the mole numbers, $U$ the internal energy, $\theta$ the contact temperature (4.8), $T^*$ the thermostatic temperature of $\mathcal{G}^*$, $\underline{A}^*$ the generalized forces of $\mathcal{G}^*$, and $\underline{\mu}^*$ the chemical potentials of $\mathcal{G}^*$. Somewhat different from (1.11) we here introduce the intensive variables of the surroundings as parameters.

The equilibrium caloric equation of state defined on the equilibrium subspace

$$T = C(\underline{a}, \underline{n}, U) \ , \qquad \frac{\partial C}{\partial U} > 0 \ , \tag{5.10}$$

gives a 1–1 map between the internal energy and the thermostatic temperature $T$ of the considered system, if it is in equilibrium. The quantities $U^*$ and $U^\theta$ are defined by

$$T^* = C(\underline{a}, \underline{n}, U^*) \ , \tag{5.11}$$

$$\theta = C(\underline{a}, \underline{n}, U^\theta) \ . \tag{5.12}$$

By projections $P^U$, $P^\theta$, and $P^*$ onto the equilibrium subspace (1.12) we introduce accompanying processes (1.13):

$$P^U Z(t) = (\underline{a}, \underline{n}, U)(t) = Z_o^U(t) \ , \tag{5.13}$$

$$P^\theta Z(t) = (\underline{a}, \underline{n}, U^\theta(t) = Z_o^\theta(t) \ , \tag{5.14}$$

$$P^* Z(t) = (\underline{a}, \underline{n}, U^*)(t) = Z_o^*(t) \ , \tag{5.15}$$

$$P^U P^U = P^U \ , \ etc. \tag{5.16}$$

## 5.3 First Law and Entropies

The first law of an open system runs as follows (2.47)

$$\dot{U}(t) = \dot{Q}(t) + \underline{A}(t) \cdot \underline{\dot{a}}(t) + \underline{h}(t) \cdot \underline{\dot{n}}^e(t) \tag{5.17}$$

Applying the projection $P^*$ we get along the accompanying process $P^* Z(t)$ :

$$\dot{U}^*(t) = \dot{Q}^*(t) + \underline{A}^*(t) \cdot \underline{\dot{a}}(t) + \underline{h}^*(t) \cdot \underline{\dot{n}}^e(t) \ , \tag{5.18}$$

$$\underline{h}^* := \frac{\partial}{\partial \underline{n}}(U^* - \underline{A}^* \cdot \underline{a}) = h^*(\underline{A}^*, \underline{n}, T^*) \ . \tag{5.19}$$

The equilibrium entropy along the $U$-projection $P^U Z(t)$ is

$$S^{\mathrm{eq}}(t) = S^{\mathrm{eq}}(\underline{a}, \underline{n}, U)(t) \ , \tag{5.20}$$

and we get Gibbs' fundamental equation as usual:

$$T\dot{S}^{\mathrm{eq}} = \dot{U} - \underline{A}^U \cdot \underline{\dot{a}} - \underline{\mu}^U \cdot \underline{\dot{n}} \ . \tag{5.21}$$

$\underline{\dot{n}}$ is the total rate of mole numbers of $\mathcal{G}$ (1.4). By other projections we get further Gibbs' equations:

$$P^\theta : \qquad \theta \dot{S}^{\mathrm{eq}} = \dot{U}^\theta - \underline{A}^\theta \cdot \underline{\dot{a}} - \underline{\mu}^\theta \cdot \underline{\dot{n}} \ , \tag{5.22}$$

$$P^* : \qquad T^* \dot{S}^{\mathrm{eq}} = \dot{U}^* - \underline{A}^* \cdot \underline{\dot{a}} - \underline{\mu}^* \cdot \underline{\dot{n}} \ . \tag{5.23}$$

These equations are projections onto equilibrium subspace of a non-equilibrium trajectory along which we have to define a non-equilibrium entropy. What can such a definition look like? Of course we have to use the contact quantities, contact temperature $\theta$, dynamic generalized forces $\underline{A}$, and the dynamic chemical potentials $\underline{\mu}$:

$$\theta \dot{S} := \dot{U} - \underline{A} \cdot \underline{\dot{a}} - \underline{\mu} \cdot \underline{\dot{n}} + \theta \sum \ . \tag{5.24}$$

The entropy production $\sum$ fixes the definition of $\dot{S}$ and determines it. Because we use small state spaces (1.11) the constitutive equations are functionals of the process history (1.10):

$$S(t) = \mathcal{S}(Z^t(\cdot)) \tag{5.25}$$

$$\underline{A}(t) = \mathcal{A}(Z^t(\cdot)) \tag{5.26}$$

$$\underline{\mu}(t) = \mathcal{M}(Z^t(\cdot)) \ . \tag{5.27}$$

*Problem 5.1* : Using the definition

$$\theta \underline{s} := \underline{h} - \underline{\mu} \tag{5.28}$$

we get by inserting the first law into the defining equations of the time rate of entropy:

$$\theta \dot{\underline{s}} := \dot{Q} + \theta \underline{s} \cdot \underline{\dot{n}}^e - \underline{\mu} \cdot \underline{\dot{n}}^i + \theta \sum \ . \tag{5.29}$$

If we consider an isolated system, we see that the last two terms of the right-hand side of (5.29) are the irreversible parts of $\theta \dot{S}$. The different projections of (5.29) are

$$P^* : \quad T^* \dot{S}^{\mathrm{eq}} = \dot{Q}^* + T^* \underline{s}^* \cdot \underline{\dot{n}}^e \ , \tag{5.30}$$

$$P^U : \quad T \dot{S}^{\mathrm{eq}} = \dot{Q} + T \underline{s}^U \cdot \underline{\dot{n}}^e \ , \tag{5.31}$$

$$P^\theta : \quad \theta \dot{S}^{\mathrm{eq}} = \dot{Q}^\theta + \theta \underline{s}^\theta \cdot \underline{\dot{n}}^e \ . \tag{5.32}$$

Of course the definition of a non-equilibrium entropy is ***not at all unique***. Another possibility is

$$T^* \dot{S}' := \dot{Q} + T^* \underline{s}^* \cdot \underline{\dot{n}}^{eq} - \underline{\mu}^\theta \cdot \underline{\dot{n}}^i + \theta \Sigma' \ . \tag{5.33}$$

The definitions (5.29) and (5.33) and other possibilities cannot be arbitrary. They have to be compatible with Clausius' inequality and the embedding axiom being introduced in the next section.

## 5.4 Clausius' Inequality and Embedding Axiom

Clausius' inequality refers to a non-equilibrium trajectory which becomes cyclic by closing it using a projection (Figure 1.8). This kind of cyclic process which contains at least one equilibrium state is marked by $+$. Clausius' inequality for open systems then writes

$$^+\oint \left[ \frac{\dot{Q}(t)}{T^*(t)} + \underline{s}^*(t) \cdot \underline{\dot{n}}^e(t) \right] dt \leq 0 \tag{5.34}$$

Notice that $T^*$ and $\underline{s}^*$ belong to the system's surroundings $\mathcal{G}^*$. Integrating $\dot{S}$ (5.29) and $\dot{S}'$ (5.33) along the considered cyclic process we get

$$^+\oint \dot{S}' dt =^+ \oint \left[ \frac{\dot{Q}}{T^*} + \underline{s}^* \cdot \underline{\dot{n}}^e \right] dt + $$

$$+^+ \oint \left[ \frac{\theta}{T^*} \Sigma' - \frac{\underline{\mu}^\theta}{\mathrm{T}^*} \cdot \underline{\dot{\mathrm{n}}}^{\mathrm{i}} \right] dt \ , \tag{5.35}$$

$$^+\oint \dot{S} dt =^+ \oint \left[ \frac{\dot{Q}}{\theta} + \underline{s} \cdot \underline{\dot{n}}^e \right] dt + $$

$$+^+ \oint \left[ \sum - \frac{\underline{\mu}}{\theta} \cdot \underline{\dot{n}}^i \right] dt \ . \tag{5.36}$$

Because definitions of non-equilibrium entropies have to be compatible with the well-known equilibrium entropy we formulate the

*Embedding Axiom* :

$$^{+}\oint \dot{S}dt =^{+} \oint \dot{S}'dt = 0 \ . \tag{5.37}$$

Notice that in general for arbitrary cyclic processes

$$\oint \dot{S}dt \neq 0 \tag{5.38}$$

is valid. Because in small state spaces the non-equilibrium entropy is as a constitutive equation a functional of the process history and not a state function. By use of (5.37) and (5.34) we get from (5.35) a dissipation inequality

$$^{+}\oint \left[\frac{\theta}{T^*}\Sigma' - \frac{\underline{\mu}^\theta}{T^*} \cdot \dot{\underline{n}}^i\right] dt \geq 0 \ . \tag{5.39}$$

The analogous inequality following from (5.36)

$$^{+}\oint \left[\sum - \frac{\underline{\mu}}{\theta} \cdot \dot{\underline{n}}^i\right] dt \geq 0 \tag{5.40}$$

does not follow from Clausius' inequality (5.34) because we need its extended formulation [24]:

$$^{+}\oint \left[\frac{\dot{Q}}{\theta} + \underline{s} \cdot \dot{\underline{n}}^e\right] dt \leq 0 \ . \tag{5.41}$$

In the special case of closed systems we can easily see that (5.41) is an extension of (5.34) because for closed systems (4.8) yields

$$^{+}\oint \frac{\dot{Q}}{\theta}dt \geq^{+} \oint \frac{\dot{Q}}{T^*}dt \ . \tag{5.42}$$

## 5.5 Non-negative Entropy Production

Consider an arbitrary process starting out from equilibrium (Figure 5.1):

$$\mathcal{F} : A(\mathrm{eq}) \rightarrow B \ . \tag{5.43}$$

In $B$ the system will be isolated and will go to equilibrium:

$$F : B \rightarrow C(\mathrm{eq}) \ , \quad \dot{Q} = 0 \ , \quad \dot{\underline{a}} = 0 \ , \quad \dot{\underline{n}}^e = \underline{0} \ . \tag{5.44}$$

The process is closed by an equilibrium trajectory

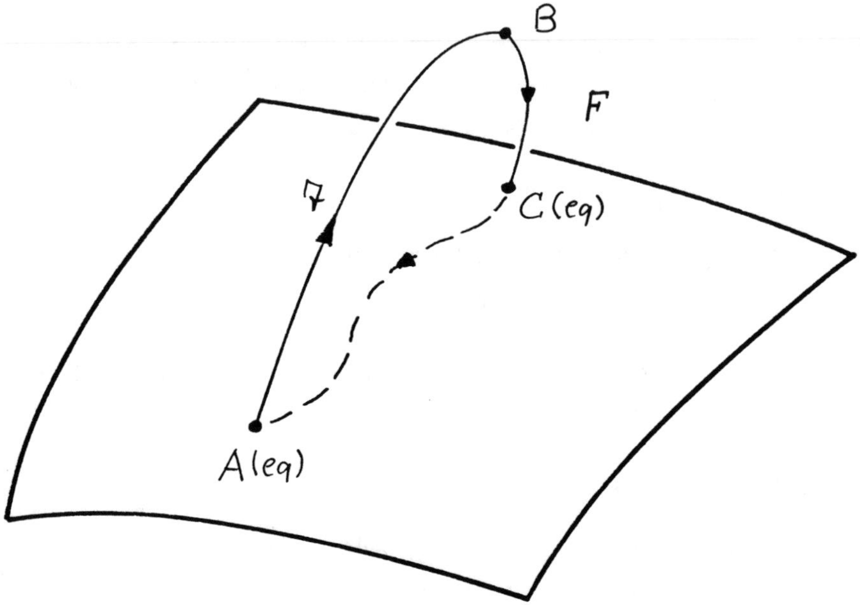


Figure 5.1: Cyclic process consisting of an arbitrary part $\mathcal{F}$ starting out from equilibrium $A(eq)$, an isolated part $F$, and an accompanying process $C$.

$$C : C(\text{eq}) \rightarrow A(\text{eq}) \ , \quad \sum = 0 \ , \quad \dot{\underline{n}}^i = \underline{0} \ . \tag{5.45}$$

*Problem 5.2* : The dissipation inequalities (5.39) and (5.40) yield for the cyclic process $\mathcal{F} \vee F \vee C$

$$\mathcal{F} \int_A^B \frac{1}{T^*} [\theta \Sigma' - \underline{\mu}^\theta \cdot \dot{\underline{n}}^i] dt \geq S'_B - S_C^{\text{eq}} \ , \tag{5.46}$$

$$\mathcal{F} \int_A^B \left[ \sum - \frac{1}{\theta} \underline{\mu} \cdot \dot{\underline{n}}^i \right] dt \geq S_B - S_C^{\text{eq}} \ . \tag{5.47}$$

The change of entropy $S_B - S_C^{\text{eq}}$ cannot be estimated because we do not know the sign of the left-hand integrals. But in the case $B \rightarrow A(\text{eq})$ (Figure 5.1), i.e. $F$ connects $A(\text{eq})$ and $B(\text{eq})$ and the system is isolated, the left-hand integrals in (5.46) and (5.47) vanish. We get

$$A(\text{eq}) \overset{\text{isol.}}{\rightarrow} C(\text{eq})$$

$$\tag{5.48}$$

$$S_C^{\text{eq}} \geq S_A^{\text{eq}} \ .$$

We now integrate the definition of $\dot{S}$ (5.29) and of $\dot{S}'$ (5.33) along $F$. By use of (5.44) this yields:

$$F \int_B^C \left[ \sum - \frac{1}{\theta} \underline{\mu} \cdot \dot{\underline{n}}^i \right] dt = S_C^{\text{eq}} - S_B \ , \tag{5.49}$$

$$F \int_B^C \frac{1}{T^*} [\theta \Sigma' - \underline{\mu}^\theta \cdot \dot{\underline{n}}^i] dt = S_C^{\text{eq}} - S'_B \ . \tag{5.50}$$

In contrast to (5.46) and (5.47) these are equations for the differences of entropy. Because $F$ is a process in an isolated system we know some more about the left-hand integrals in (5.49) and (5.50) than about those in (5.46) and (5.47): Because we are in small state spaces these integrals are functionals of the process history up to the state

$$Z^B = (\underline{a}^B, \underline{n}^B, U^B, \theta^B, \ldots; T_B^*, \underline{A}_B^*, \underline{\mu}_B^*) \ . \tag{5.51}$$

Because of the isolation along $F$ after passing $B$ the history influences the value of the integrals only until the time $t_B$ from which the isolation begins. Along $F$ the integral depends only on the differences between $B$ and $C$ of those variables which change despite the isolation. These variables are the contact temperature $\theta$, or according to (5.12) $U^\theta$, and the mole numbers due to chemical reactions. Therefore we get introducing the abbreviations

$$U^\theta(t) - U_C =: \xi(t) \ , \qquad \underline{n}(t) - \underline{n}_C =: \underline{\eta}(t) \ , \tag{5.52}$$

$$F\int_B^C \left[\Sigma - \frac{1}{\theta}\underline{\mu} \cdot \dot{\underline{n}}^i\right] dt =$$

$$= -F(\xi(t_B) \ , \ \underline{\eta}(t_B) \ , \ldots, \ Z^{t_B}(\cdot)) \ . \tag{5.53}$$

Here $F$ is a functional of the process history up to time $t_B$. Because $Z^{t_B}(\cdot)$ is the only parameter along $F$ the total differential of $F$ along $F$ writes:

$$DF = \frac{\partial F}{\partial \xi}(\xi, \underline{\eta}, \ldots, Z^{t_B}(\cdot))\dot{\xi} +$$

$$+ \frac{\partial F}{\partial \underline{\eta}}(\xi, \underline{\eta}, \ldots, Z^{t_B}(\cdot)) \cdot \dot{\underline{\eta}} + \ldots \ . \tag{5.54}$$

Choosing along $F$

$$\left(\Sigma - \frac{1}{\theta}\underline{\mu} \cdot \dot{\underline{n}}^i\right)(t) =$$

$$= DF(\xi(t), \underline{\eta}(t), \ldots, Z^{t_B}(\cdot)) \tag{5.55}$$

we get by integration along $F$

$$F\int_B^C (\Sigma - \frac{1}{\theta}\underline{\mu} \cdot \dot{\underline{n}}^i) dt =$$

$$= F(0, \underline{0}, \ldots, Z^{t_B}(\cdot)) - F(\xi(t_B), \underline{\eta}(t_B), \ldots, Z^{t_B}(\cdot)) \ . \tag{5.56}$$

Up to here the functional $F$ is arbitrary. Now we demand

$$F(0, \underline{0}, \ldots, Z^{t_B}(\cdot)) \equiv 0 \ , \tag{5.57}$$

and consequently (5.53) is satisfied.

The "process speed" along $F$ is determined by constitutive equations

$$\dot{\xi}(t) = X(\xi(t), \underline{\eta}(t), \ldots, Z^{t_B}(\cdot)) \ , \tag{5.58}$$

$$\dot{\underline{\eta}}(t) = E(\xi(t), \underline{\eta}(t), \ldots, Z^{t_B}(\cdot)) \ . \tag{5.59}$$

In another diction (5.58) and (5.59) are relaxation rate equations in an isolated system of a process starting out from $B$ at time $t_B$. For enforcing non-negative entropy production we choose

$$DF \geq 0 \ , \tag{5.60}$$

which can be achieved by

$$\mathrm{sgn}\frac{\partial F}{\partial \xi} \stackrel{!}{=} \mathrm{sgn}\dot{\xi} \ , \qquad \mathrm{sgn}\frac{\partial F}{\partial \underline{\eta}} \stackrel{!}{=} \mathrm{sgn}\dot{\underline{\eta}} \ . \tag{5.61}$$

This is a very weak condition to $F$ because it only concerns signs. In contrast to $F$ the functionals $X$ and $E$ are determined by the material. A possible dependence of $F$ from $\xi$ is sketched in Figure 5.2.

Finally we put together the results. According to (5.55) and (5.60) we have

$$(\Sigma - \frac{1}{\theta}\underline{\mu} \cdot \dot{\underline{n}}^{\mathrm{i}})(t) \geq 0 \ , \ \text{along } F \ . \tag{5.62}$$

Consequently according to (5.49) and (5.50) we get

$$\begin{aligned} & B \stackrel{\text{isol.}}{\rightarrow} C(\text{eq}) \\ & S_C^{\text{eq}} \geq S_B \\ & S_C^{\text{eq}} \geq S'_B \end{aligned} \tag{5.63}$$

and beyond that also

$$\begin{aligned} & \dot{S}(t) \geq 0 \\ & \dot{S}'(t) \geq 0 \end{aligned} \tag{5.64}$$

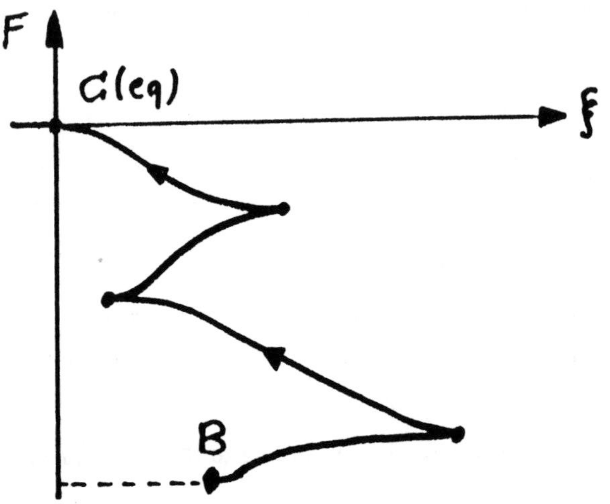


Figure 5.2: Graph showing the behavior of $F$ taking into consideration (5.60) and (5.61).

along $F$. According to (5.55) the entropy production $\sum$ and the chemical potentials $\underline{\mu}$ are also functionals of the process history

$$\sum(t) = \sum(Z^t(\cdot)) \ , \tag{5.65}$$

$$\underline{\mu}(t) = \underline{\mu}(Z^t(\cdot)) \ . \tag{5.66}$$

These equations show that the entropy production and the chemical potentials do not depend on the time derivatives of the state space variables and consequently they are independent of the process direction in state space. But along $F$, $DF$ is definite and because of its independence of the process direction, it is definite in every direction. So we get:

$$(\sum - \frac{1}{\theta}\underline{\mu} \cdot \underline{\dot{n}}^i)(t) \geq 0 \quad \text{holds generally.} \tag{5.67}$$

(5.29) and (5.33) yield

$$\dot{S} \geq \frac{\dot{Q}}{\theta} + \underline{s} \cdot \underline{\dot{n}}^e \ , \tag{5.68}$$

$$\dot{S}' \geq \frac{\dot{Q}}{T^*} + \underline{s}^* \cdot \underline{\dot{n}}^e \ . \tag{5.69}$$

Consequently we proved a statement which is used in continuum thermodynamics as a formulation of the second law without any remark about its connection to Clausius' inequality [23]: It is always possible, but not uniquely, to give a material dependent definition of entropy production so that it will be definite.

# Chapter 6

# Continuum Thermodynamics

Up to here we dealt with discrete systems which are characterized by exchange quantities and a (non-equilibrium) state space. If we divide a system into sufficiently small subsystems, each of them described as a discrete system, we get the so-called field formulation of thermodynamics which uses fields of thermodynamical quantities. Most of these fields contain no difficulty with regard to their definition and interpretation. But the fields of specific entropy and temperature have to be considered in more detail, because up to now contact temperature is only defined for discrete systems. Therefore the question arises:

## 6.1 How to Achieve Field Formulation

We will consider three examples to see the translation from discrete systems to field formulation: the balance equations of mass, momentum, and entropy.

### 6.1.1 Partial Mass Balance

We consider a discrete open system $\mathcal{G}$ containing $K$ components reacting with each other [25]. The mass balance of the $\alpha$-th component is

$$^{\alpha}\dot{M}(\mathcal{G}(t)) = {^{\alpha}\dot{M}_{\text{chem}}} + {^{\alpha}\dot{M}(\partial\mathcal{G}(t))} \ . \tag{6.1}$$

The right-hand terms are the rate of mass of the $\alpha$-th component due to chemical reaction and due to exchange through the partition $\partial\mathcal{G}$ between $\mathcal{G}$ and its environment. Summing up the components we get

$$\dot{M}(\partial\mathcal{G}(t)) = \sum_{\alpha} {}^{\alpha}\dot{M}(\partial\mathcal{G}(t)) \ , \tag{6.2}$$

$$0 = \sum_{\alpha} {}^{\alpha}\dot{M}_{\text{chem}} \ . \tag{6.3}$$

Equation (6.3) represents mass conservation.

Dividing $\partial\mathcal{G}$ into subsurfaces we get a surface field formulation on $\partial\mathcal{G}$:

$${}^{\alpha}\dot{M}(\partial\mathcal{G}(t)) \to - \oint_{\partial\mathcal{G}(t)} \underline{\dot{m}}^{\alpha}(\underline{x}, t) \cdot d\underline{f} \ . \tag{6.4}$$

Here $\underline{\dot{m}}^{\alpha}$ is the *partial mass flux density* of the $\alpha$-th component. Dividing $\mathcal{G}$ into subsystems

$$\mathcal{G} = \cup_i \mathcal{G}_i \ , \qquad \mathcal{G}_i \cap \mathcal{G}_k = \emptyset \ , \qquad i \neq k \ , \tag{6.5}$$

$$V_i = \text{vol}(\mathcal{G}_i) \ , \tag{6.6}$$

we get

$$\frac{{}^{\alpha}\dot{M}(\mathcal{G}_i(t))}{V_i} = \frac{{}^{\alpha}\dot{M}_{\text{chem}}}{V_i} - \frac{1}{V_i} \oint_{\partial\mathcal{G}_i} \underline{\dot{m}}^{\alpha} \cdot d\underline{f} \ . \tag{6.7}$$

The left-hand side writes:

$$\frac{{}^{\alpha}\dot{M}(\mathcal{G}_i)}{V_i} = \frac{d}{dt} \frac{{}^{\alpha}M(\mathcal{G}_i)}{V_i} + \frac{{}^{\alpha}M(\mathcal{G}_i)}{V_i} \frac{1}{V_i} \dot{V}_i \ . \tag{6.8}$$

By use of Reynolds' transport theorem

$$\frac{d}{dt} \int_{\mathcal{G}(t)} \phi dV = \int_{\mathcal{G}(t)} (\dot{\phi} + \phi \nabla \cdot \underline{w}) dV \ , \tag{6.9}$$

$$\frac{d}{dt} = \frac{\partial}{\partial t} + \underline{w} \cdot \frac{\partial}{\partial \underline{x}} \ , \tag{6.10}$$

we get

$$\dot{V}_i = \frac{d}{dt} \int_{\mathcal{G}_i} dV = \oint_{\partial\mathcal{G}_i} \underline{w} \cdot d\underline{f} \ . \tag{6.11}$$

Introducing arbitrary small subsystems we get the *partial mass density*

$$\lim_{\mathcal{G}_i \to 0} \frac{{}^{\alpha}M(\mathcal{G}_i)}{V_i} =: \rho^{\alpha}(\underline{x}, t) \ , \tag{6.12}$$

the *partial mass production*

$$\lim_{\mathcal{G}_i \to 0} \frac{{}^{\alpha}\dot{M}_{\text{chem}}}{V_i} =: \tau^{\alpha}(\underline{x}, t) \ , \tag{6.13}$$

the divergence of the *velocity of the map* $\mathcal{G}(t_o) \to \mathcal{G}(t)$

$$\lim_{\mathcal{G}_i \to 0} \frac{1}{V_i} \oint_{\partial \mathcal{G}_i} \underline{w} \cdot d\underline{f} =: \nabla \cdot \underline{w} \ , \tag{6.14}$$

the *partial mass flux density*

$$\lim_{\mathcal{G}_i \to 0} \frac{1}{V_i} \oint_{\partial \mathcal{G}_i} \underline{\dot{m}}^{\alpha} \cdot d\underline{f} =: \nabla \cdot \underline{\dot{m}}^{\alpha} \ . \tag{6.15}$$

Introducing the quantities into (6.8) we have

$$\dot{\rho}^{\alpha} + \rho^{\alpha} \nabla \cdot \underline{w} = \tau^{\alpha} - \nabla \cdot \underline{\dot{m}}^{\alpha} \ . \tag{6.16}$$

The *partial material velocity* is defined by

$$\rho^{\alpha} \underline{v}^{\alpha} := \underline{\dot{m}}^{\alpha} + \rho^{\alpha} \underline{w} \ , \tag{6.17}$$

and (6.16) yields the *partial mass balance*:

$$\frac{\partial}{\partial t} \rho^{\alpha} = -\nabla \cdot (\underline{\dot{m}}^{\alpha} + \rho^{\alpha} \underline{w}) + \tau^{\alpha} \ . \tag{6.18}$$

Summing up the components and defining the *mass density*

$$\rho := \sum_{\alpha} \rho^{\alpha} \ , \tag{6.19}$$

and the *material velocity*

$$\rho \underline{v} := \sum_{\alpha} \rho^{\alpha} \underline{v}^{\alpha} \tag{6.20}$$

we get the *mass balance* equation,

$$\frac{\partial}{\partial t} \rho = -\nabla \cdot \rho \underline{v} \ , \tag{6.21}$$

and we can formulate the

*Axiom of Mixtures* : Summing up the components of a partial balance equation the sum looks like a balance equation of a 1-component system.

### 6.1.2 Partial Momentum Balance

The partial momentum of a discrete system is defined by

$$^{\alpha}\underline{P}(\mathcal{G}(t)) := \int_{\mathcal{G}(t)} \rho^{\alpha}(\underline{x},t)\underline{v}^{\alpha}(\underline{x},t)dV \ . \tag{6.22}$$

The balance of momentum writes

$$^{\alpha}\dot{\underline{P}}(\mathcal{G}) = {}^{\alpha}\underline{K} + {}^{\alpha}\dot{\underline{P}}(\partial\mathcal{G}) + {}^{\alpha}\underline{M}(\mathcal{G}) \ . \tag{6.23}$$

Here $^{\alpha}\underline{K}$ are the forces, $^{\alpha}\dot{\underline{P}}(\partial\mathcal{G})$ the supply due to the opening of the system, and $^{\alpha}\underline{M}(\mathcal{G})$ the momentum interaction. The forces can be split into volume forces $^{\alpha}\underline{F}$ and surface forces $^{\alpha}\underline{t}$

$$^{\alpha}\underline{K} = {}^{\alpha}\underline{F} + {}^{\alpha}\underline{t} \ . \tag{6.24}$$

The division procedure into subsystems yields the *partial acceleration*

$$^{\alpha}\underline{F}(\mathcal{G}_i) \to \rho^{\alpha}(\underline{x},t)\underline{f}^{\alpha}(\underline{x},t) \ , \tag{6.25}$$

and the volume force

$$\underline{F}^{\alpha}(\mathcal{G}) = \int_{\mathcal{G}(t)} \rho^{\alpha}\underline{f}^{\alpha}dV \ . \tag{6.26}$$

The surface forces $^{\alpha}\underline{t}$ are contact quantities as the generalized forces in (4.80). Their defining inequality is

$$({}^{\alpha}\underline{t}^{\text{eq}} - {}^{\alpha}\underline{t}) \cdot \dot{\underline{g}}^{\alpha} \geq 0 \ , \tag{6.27}$$

and the indicator $\dot{\underline{g}}^{\alpha}$ is the time rate of the elongation of a spring fastened at an $\alpha$-impermeable piston on the surface $\partial\mathcal{G}$ (Figure 6.1). By this procedure we get the surface function of the contact force $^{\alpha}\underline{t}(\underline{x},\underline{n},t)$. The total surface force is

$$^{\alpha}\underline{t}(\partial\mathcal{G}) = \oint_{\partial\mathcal{G}} {}^{\alpha}\underline{t}(\underline{x},\underline{n},t)df = -\oint_{\partial\mathcal{G}} \underline{\underline{P}}^{\alpha}(\underline{x},t)\cdot d\underline{f} \ . \tag{6.28}$$

The second equation is obtained by the usual reasoning using Cauchy's tetrahedron argumentation. $\underline{\underline{P}}^{\alpha}$ is the *partial pressure tensor* field.

The supply in (6.23) is defined by

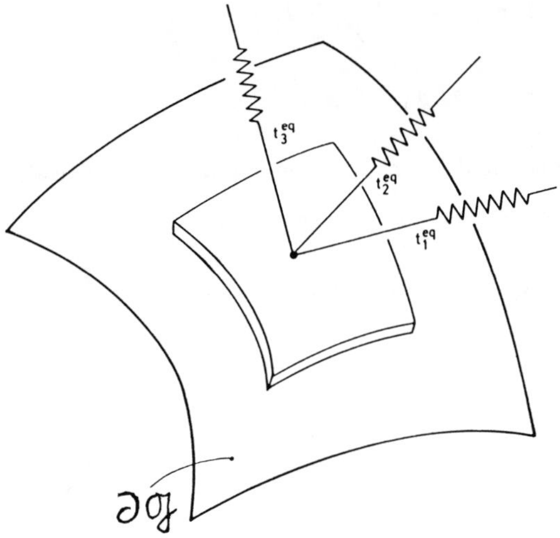


Figure 6.1: An $\alpha$-impermeable piston on $\partial \mathcal{G}$ is in relative rest to $\partial \mathcal{G}$. The springs indicate the surface contact force.

$$ {}^{\alpha}\dot{\underline{P}}(\partial\mathcal{G}) := -\oint_{\partial\mathcal{G}(t)} \underline{v\dot{m}^{\alpha}} \cdot d\underline{f} \ . \tag{6.29} $$

The momentum conservation is given by

$$ \sum_{\alpha} {}^{\alpha}\underline{M}(\mathcal{G}) = \underline{0} \ . \tag{6.30} $$

Inserting (6.22), (6.26), (6.28), and (6.29) into (6.23) we get the *global partial momentum balance* equation:

$$ \frac{d}{dt}\int_{\mathcal{G}} \rho^{\alpha}\underline{v}^{\alpha} dV = \int_{\mathcal{G}} \rho^{\alpha}\underline{f}^{\alpha} dV - \oint_{\partial\mathcal{G}} \underline{\underline{P}}^{\alpha} \cdot d\underline{f} - $$

$$ -\oint_{\partial\mathcal{G}} \rho^{\alpha}\underline{v}(\underline{v}^{\alpha} - \underline{w}) \cdot d\underline{f} + {}^{\alpha}\underline{M}(\mathcal{G}) \ . \tag{6.31} $$

*Problem 6.1* : Using Reynolds' transport theorem we get two equivalent formulations of the *local partial momentum balance* equation:

$$ (\rho^{\alpha}\dot{\underline{v}}^{\alpha}) + \rho^{\alpha}\underline{v}^{\alpha}\nabla \cdot \underline{w} + \nabla \cdot \underline{\underline{P}}^{\alpha} = \rho^{\alpha}\underline{f}^{\alpha} - $$

$$ -\nabla \cdot \underline{\dot{m}}^{\alpha}\underline{v}^{\alpha} + \underline{m}^{\alpha} \Leftrightarrow \tag{6.32} $$

$$ \frac{\partial}{\partial t}(\rho^{\alpha}\underline{v}^{\alpha}) + \nabla \cdot (\rho^{\alpha}\underline{v}^{\alpha}\underline{v}^{\alpha} + \underline{\underline{P}}^{\alpha}) = \rho^{\alpha}\underline{f}^{\alpha} + \underline{m}^{\alpha} \ . \tag{6.33} $$

*Problem 6.2* : Using the axiom of mixtures the partial quantities and the quantities belonging to the mixture as a whole are related by:

$$ \underline{P}(\mathcal{G}) = \sum_{\alpha} {}^{\alpha}\underline{P}(\mathcal{G}) \ , \tag{6.34} $$

$$ \dot{\underline{P}}(\partial\mathcal{G}) = \sum_{\alpha} {}^{\alpha}\dot{\underline{P}}(\partial\mathcal{G}) - \oint_{\partial\mathcal{G}} \sum_{\alpha} \rho^{\alpha}\underline{v}^{\alpha}(\underline{v} - \underline{v}^{\alpha}) \cdot d\underline{f} \ , \tag{6.35} $$

$$ \underline{t} = \sum_{\alpha} {}^{\alpha}\underline{t} - \{\dot{\underline{P}}(\partial\mathcal{G}) - \sum_{\alpha} {}^{\alpha}\dot{\underline{P}}(\partial\mathcal{G})\} \ , \tag{6.36} $$

$$ \underline{\underline{P}} = \sum_{\alpha} \{\underline{\underline{P}}^{\alpha} + \rho^{\alpha}(\underline{v}^{\alpha} - \underline{v})(\underline{v}^{\alpha} - \underline{v})\} \ . \tag{6.37} $$

### 6.1.3 Entropy Balance

In Section 5.3 time rates of non-equilibrium entropies were introduced by (5.29) and (5.33). These entropies belong to the mixture as a whole. We could write down also equations for time rates of partial entropies, but partial entropy productions are in general not definite. Therefore we will not use such equations.

Starting out with heat exchanges we state heat exchanges through subsurfaces are additive:

$$\dot{Q}(\partial \mathcal{G}) = \sum_i \dot{Q}(\partial \mathcal{G}_i) = -\oint_{\partial \mathcal{G}} \underline{q} \cdot d\underline{f} \ . \tag{6.38}$$

In contrast to heat exchanges entropy flux densities are not additive. We will prove this statement in two steps:

*Proposition 6.1* The dissipation inequality (4.65) yields (4.8)

$$\dot{Q}\left(\frac{1}{\theta} - \frac{1}{T^*}\right) \geq 0 \ . \tag{6.39}$$

*Proof*: The dynamic chemical potentials $\underline{\mu}$ are defined by (4.17)

$$T^* = \theta : \ \underline{\dot{n}}^e \cdot (\underline{\mu}^* - \underline{\mu}) \geq 0 \ . \tag{6.40}$$

Because $\dot{Q}$ is independent of $\underline{\mu}$ and $\underline{\mu}^*$ by definition we choose

$$\underline{\mu}^*/T^* = \underline{\mu}/\theta \tag{6.41}$$

Then (4.65) yields (6.39). □

*Proposition 6.2*: If $\{\partial \mathcal{G}_i\}$ is a division of $\mathcal{G}$,

$$\cup_i \partial \mathcal{G}_i = \partial \mathcal{G} \ , \quad \partial \mathcal{G}_j \cap \partial \mathcal{G}_k = \emptyset \ , \quad j \neq k \ ,$$

the inequality

$$\sum_i \frac{\dot{Q}(\partial \mathcal{G}_i)}{\theta(\partial \mathcal{G}_i)} \geq \frac{\dot{Q}(\partial \mathcal{G})}{\theta(\partial \mathcal{G})} \tag{6.42}$$

holds.

*Proof*: Define subsurfaces by

$$\dot{Q}(\partial \mathcal{G}_i^+) \geq 0 \ , \qquad \dot{Q}(\partial \mathcal{G}_i^-) < 0 \ . \tag{6.43}$$

Then we get

$$\sum_i \frac{\dot{Q}(\partial \mathcal{G}_i)}{\theta(\partial \mathcal{G}_i)} = \sum_j \frac{\dot{Q}(\partial \mathcal{G}_j^+)}{\theta(\partial \mathcal{G}_j^+)} + \sum_k \frac{\dot{Q}(\partial \mathcal{G}_k^-)}{\theta(\partial \mathcal{G}_k^-)} \ . \tag{6.44}$$

Using the mean value theorem (6.44) yields

$$\sum_i \frac{\dot{Q}(\partial \mathcal{G}_i)}{\theta(\partial \mathcal{G}_i)} = \frac{\dot{Q}^+}{\theta^+} + \frac{\dot{Q}^-}{\theta^-} \tag{6.45}$$

Consequently the surface $\partial \mathcal{G}$ can be divided into two subsurfaces $\partial \mathcal{G}^+$ and $\partial \mathcal{G}^-$. The heat exchange through $\partial \mathcal{G}^+$ is $\dot{Q}^+ \geq 0$, that through $\partial \mathcal{G}^-$ is $\dot{Q}^- < 0$. The mean value of the contact temperatures are $\theta^+$ on $\partial \mathcal{G}^+$ and $\theta^-$ on $\partial \mathcal{G}^-$. Because the considered system is placed in an equilibrium environment of thermostatic temperature $T$ we have on the subsurfaces

$$\partial \mathcal{G}_j^+ : \ \dot{Q}_j^+ \left( \frac{1}{\theta_j^+} - \frac{1}{T} \right) \geq 0 \ , \tag{6.46}$$

$$\partial \mathcal{G}_k^- : \ \dot{Q}_k^- \left( \frac{1}{\theta_k^-} - \frac{1}{T} \right) \geq 0 \ . \tag{6.47}$$

Summing up gives

$$\partial \mathcal{G}^+ : \ \dot{Q}^+ \left( \frac{1}{\theta^+} - \frac{1}{T} \right) \geq 0 \ , \tag{6.48}$$

$$\partial \mathcal{G}^- : \dot{Q}^- \left( \frac{1}{\theta^-} - \frac{1}{T} \right) \geq 0 \ . \tag{6.49}$$

Consequently the mean value $\theta^+$ and $\theta^-$ are the contact temperatures of $\partial \mathcal{G}^+$ and $\partial \mathcal{G}^-$. The contact temperature $\theta$ of $\partial \mathcal{G}$ is defined by

$$T = \theta : \dot{Q}^+ + \dot{Q}^- = \dot{Q} = 0 \ . \tag{6.50}$$

Inserting this into (6.48) and (6.49) we get

$$\theta^+ \leq \theta \leq \theta^- \tag{6.51}$$

which is a remarkable result. Using (6.51) the following inequalities are valid

$$\frac{\dot{Q}^+}{\theta^+} \geq \frac{\dot{Q}^+}{\theta} \ , \quad \frac{\dot{Q}^-}{\theta^-} \geq \frac{\dot{Q}^-}{\theta} \ . \tag{6.52}$$

Adding both the inequalitites we get by use of (6.45) the proposition. □

If we would interpret $\dot{Q}(\partial \mathcal{G}_i)/\theta(\partial \mathcal{G}_i)$ as entropy flux through $\partial \mathcal{G}$, (6.42) shows entropy fluxes are not additive.

We now translate the dissipation inequalities (5.68) and (5.69):

$$S = \int_{\mathcal{G}(t)} (\sum_\alpha \rho^\alpha \hat{s}^\alpha) dV \ , \tag{6.53}$$

$$\frac{\dot{Q}}{\theta} = - \oint_{\partial \mathcal{G}} \underline{\phi} \cdot d\underline{f} \ , \tag{6.54}$$

$$\underline{s} \cdot \underline{\dot{n}}^e = - \oint_{\partial \mathcal{G}} (\sum_\alpha \hat{s}^\alpha \underline{\dot{m}}^\alpha) \cdot d\underline{f} \ . \tag{6.55}$$

Here $\hat{s}^\alpha$ is the specific partial entropy of the $\alpha$-th component, $\underline{\phi}$ the entropy flux density. (5.68) yields the global entropy balance

$$\frac{d}{dt} \int_{\mathcal{G}} (\sum_\alpha \rho^\alpha \hat{s}^\alpha) dV + \oint_{\partial \mathcal{G}} (\underline{\phi} + \sum_\alpha \hat{s}^\alpha \underline{\dot{m}}^\alpha) \cdot d\underline{f} \geq 0 \ , \tag{6.56}$$

and (5.69) an inequality looking totally equal:

$$\frac{d}{dt} \int_{\mathcal{G}} (\sum_\alpha \rho^\alpha \hat{s}'^\alpha) dV + \oint_{\partial \mathcal{G}} (\underline{\phi}^* + \sum_\alpha \hat{s}'^\alpha \underline{\dot{m}}^\alpha) \cdot d\underline{f} \geq 0 \ . \tag{6.57}$$

Because $\hat{s}^\alpha$, $\underline{\phi}$, $\hat{s}'^\alpha$, and $\underline{\phi}^*$ are given by constitutive equations both the dissipation inequalities (6.56) and (6.57) are equivalent. The different possibilities of defining entropy have no influence on the formal shape of the dissipation inequality. Therefore in continuum thermodynamics we can use dissipative inequalities without a specification of entropy.

The local formulation of (6.56) writes

$$\frac{\partial}{\partial t} \sum_\alpha \rho^\alpha \hat{s}^\alpha + \nabla \cdot (\underline{\phi} + \sum_\alpha \rho^\alpha \hat{s}^\alpha \underline{v}^\alpha) \geq 0 \ . \tag{6.58}$$

## 6.2 Balance Equations

We now write down the balance equation without further comment [25]:

### 6.2.1 Mass

$$\frac{\partial \rho^\alpha}{\partial t} = -\nabla \cdot \rho^\alpha \underline{v}^\alpha + \tau^\alpha \ , \qquad \sum_\alpha \tau^\alpha = 0 \ . \tag{6.59}$$

### 6.2.2 Momentum

$$\frac{\partial}{\partial t}(\rho^\alpha \underline{v}^\alpha) + \nabla \cdot (\underline{\underline{P}}^\alpha + \rho^\alpha \underline{v}^\alpha \underline{v}^\alpha) = \rho^\alpha \underline{f}^\alpha + \underline{m}^\alpha \ ,$$

$$\sum_\alpha \underline{m}^\alpha = \underline{0} \ . \tag{6.60}$$

### 6.2.3 Kinetic Energy

$$\frac{1}{2}\frac{\partial}{\partial t}[\rho^\alpha (\underline{v}^\alpha)^2] - \nabla \cdot (\underline{\underline{P}}^\alpha \cdot \underline{v}^\alpha - \frac{1}{2}\rho^\alpha (\underline{v}^\alpha)^2 \underline{v}^\alpha) =$$

$$= \underline{\underline{P}}^\alpha : \nabla \underline{v}^\alpha + \rho^\alpha \underline{f}^\alpha \cdot \underline{v}^\alpha - \frac{1}{2}\tau^\alpha (\underline{v}^\alpha)^2 +$$

$$+ \underline{m}^\alpha \cdot \underline{v}^\alpha \ . \tag{6.61}$$

### 6.2.4 Internal Energy

$$\frac{\partial}{\partial t}\rho^\alpha \epsilon^\alpha + \nabla \cdot (\underline{q} + \rho^\alpha \epsilon^\alpha \underline{v}^\alpha) = -\underline{\underline{P}}^\alpha : \nabla \underline{v} + \omega^\alpha \tag{6.62}$$

### 6.2.5 Total Energy

$$e^\alpha := \epsilon^\alpha + \frac{1}{2}(\underline{v}^\alpha)^2 \ , \tag{6.63}$$

$$\frac{\partial}{\partial t}(\rho^\alpha e^\alpha) + \nabla \cdot (\underline{q}^\alpha + \underline{\underline{P}}^\alpha \cdot \underline{v}^\alpha + \rho^\alpha e^\alpha \underline{v}^\alpha) =$$

$$= \rho^\alpha \underline{f}^\alpha \cdot \underline{v}^\alpha + \ell^\alpha \ , \tag{6.64}$$

$$\ell^\alpha := \omega^\alpha + \frac{1}{2}\tau^\alpha (\underline{v}^\alpha)^2 - \underline{m}^\alpha \cdot \underline{v}^\alpha \ ,$$

$$\sum_\alpha \ell^\alpha = 0 \ . \tag{6.65}$$

### 6.2.6 Relations between Partial and Mixture Quantities

$$\rho e := \sum_\alpha \rho^\alpha e^\alpha \tag{6.66}$$

$$\underline{\underline{P}} = \sum_\alpha \{\underline{\underline{P}}^\alpha + \rho^\alpha(\underline{v}^\alpha - \underline{v})(\underline{v}^\alpha - \underline{v})\} \tag{6.67}$$

$$\rho\epsilon = \sum_\alpha \rho^\alpha\{\epsilon^\alpha + \frac{1}{2}(\underline{v}^\alpha - \underline{v})(\underline{v}^\alpha + \underline{v})\} \tag{6.68}$$

$$\underline{q} = \sum_\alpha \{\underline{q}^\alpha + \rho^\alpha(\epsilon^\alpha + \frac{1}{2}(\underline{v}^\alpha - \underline{v})^2)(\underline{v}^\alpha - \underline{v}) + \\ + \underline{\underline{P}}^\alpha \cdot (\underline{v}^\alpha - \underline{v})\} \tag{6.69}$$

### 6.2.7 Dissipation Inequality

$$\frac{\partial}{\partial t}\sum_\alpha \rho^\alpha \hat{s}^\alpha + \nabla \cdot (\underline{\phi} + \sum_\alpha \rho^\alpha \hat{s}^\alpha \underline{v}^\alpha) \geq 0 \ . \tag{6.70}$$

## 6.3 Constitutive Equations

We consider especially a 1-component system. There are five balance equations and the dissipation inequality for 23 fields:

| number of fields | | number of equations | |
|---|---|---|---|
| 4 | $\dot{\rho} = -\rho\nabla\cdot\underline{v}$ | 1 | (6.71) |
| 9 | $\rho\underline{\dot{v}} = -\nabla\cdot\underline{\underline{P}} + \rho\underline{f}$ | 3 | (6.72) |
| 5 | $\rho\dot{\epsilon} = -\nabla\cdot\underline{q} - \underline{\underline{P}} : \nabla\underline{v} + r$ | 1 | (6.73) |
| 5 | $\rho\dot{s} \geq -\nabla\cdot\underline{\phi} + \psi$ | (1) | (6.74) |
| 23 | | 5 | |

We now have to determine what the fields are we are looking for. In literature there are two possibilities [3,26], the so-called 5-field theory and the 13-field theory.

### 6.3.1 Five-Field Theory

Five-field theories are characterized by looking for five fields, namely $(\rho, \underline{v}, \epsilon)(\underline{x}, t)$ or $(\rho, \underline{v}, \theta)(\underline{x}, t)$. They are called the ***basic fields***. Because we are looking for 5 fields and 3 fields $\underline{f}(\underline{x}, t)$ are given we need an additional 15 fields which have to be given by constitutive equations $(5 + 3 + 15 = 23)$:

| | number of the fields |
|---|---|
| $\underline{\underline{P}}(\underline{x}, t)$ | 6 |
| $\underline{q}(\underline{x}, t)$ | 3 |
| $r(\underline{x}, t)$ | 1 |
| $s(\underline{x}, t)$ | 1 |
| $\underline{\phi}(\underline{x}, t)$ | 3 |
| $\psi(\underline{x}, t)$ | 1 |
| | 15 |

The other possibility is a

### 6.3.2 Thirteen-Field Theory

Here we are looking for 13 basic fields [27]: $(\rho, \underline{v}, \epsilon, \underline{\underline{P}}^0, \underline{q})(\underline{x}, t)$ $(1 + 3 + 1 + 5 + 3 = 13)$, and $\underline{\underline{P}}^0$ is defined by

$$\underline{\underline{P}}^0 := \underline{\underline{P}} - \left(\frac{1}{3} Tr \underline{\underline{P}}\right) \underline{\underline{1}} \ , \tag{6.75}$$

the traceless part of the pressure tensor. In the 5-field theory there are 5 balances, here in the 13-field theory we need beyond these 5 balances 8 more balances for the additional fields $\underline{\underline{P}}^0$ and $\underline{q}$:

$$\dot{\underline{\underline{P}}}^0 + \nabla \cdot \underline{\underline{\underline{Q}}}^0 = \underline{\underline{F}}^0 \ , \tag{6.76}$$

$$\dot{\underline{q}} + \nabla \cdot \underline{\underline{L}} = \underline{G} \ . \tag{6.77}$$

By introducing these 8 more balance equations we also introduce quantities which are determined by constitutive equations:

| | number of fields |
|---|---|
| $P(\underline{x},t) := \frac{1}{3} Tr \underline{\underline{P}}(\underline{x},t)$ | 1 |
| $\underline{\underline{\underline{Q}}}^0(\underline{x},t)$ | 15 |
| $\underline{\underline{F}}^0(\underline{x},t)$ | 5 |
| $\underline{\underline{L}}(\underline{x},t)$ | 9 |
| $\underline{G}(\underline{x},t)$ | 3 |
| $r(\underline{x},t)$ | 1 |
| $s(\underline{x},t)$ | 1 |
| $\underline{\phi}(\underline{x},t)$ | 3 |
| $\psi(\underline{x},t)$ | 1 |
| | 39 |

Comparing the number of quantities determined by constitutive equations in 5-field theories with those of 13-field theories we see it arises from 15 to 39. So the situation is even worse in 13-field theories, but because the number of basic fields is raised from 5 to 13 the constitutive equations may be only chosen approximately. Such a choice may be

$$\underline{\underline{L}} = -\frac{1}{\tau}\underline{\underline{\kappa}}\theta \ , \qquad \underline{G} = \frac{1}{\tau}\underline{q} \ , \tag{6.78}$$

so that (6.77) yields

$$\tau\dot{\underline{q}} - \nabla\cdot\underline{\underline{\kappa}}\theta = \underline{q} \ . \tag{6.79}$$

This procedure is not at all a systematic treatment of the additional balance equations (6.76) and (6.77), but it may help to get better results than using the 5-field theory.

### 6.3.3 State Space

The basic fields are independent of each other not only in 5-field theories but also in 13-field theories. But they have nothing to do with the thermodynamical state space. As an example we mention the constitutive equation of the heat flux density

$$\underline{q} = \underline{Q}(\nabla\theta, \ldots) \tag{6.80}$$

in which $\nabla\theta$ appears, but $\nabla\theta$ is not included in the state space $(\rho, \underline{v}, \epsilon)(\underline{x}, t)$. Therefore it is obvious we need the notion of the state space also in continuum thermodynamics.

After having chosen the state space the symbols $\dot{}$ and $\nabla$ in the balance equations get a meaning: They have to be performed by the chain rule, and we can see the *balance equations determine directional derivatives in state space.* Now the questions arise: Are all directional derivatives allowed, or are they restricted by the dissipation inequality? How have we to exploit the dissipation inequality? How to establish constitutive equations which exceed a simple ansatz being compatible with the dissipation inequality? These questions lead up to formulating material axioms.

## 6.4 Material Axioms

The equation "how to get constitutive equations?" can be answered in two ways: We can make *ansatzes* compatible with the second law. Examples for such ansatzes are:

Gibbs fundamental equation

thermal equation of state

rate equations from entropy production

linear phenomenological coefficients

dissipative potentials

yield functions

multi-yield surfaces

non-identified internal variables

The other way is to formulate *material axioms*. Instead of a special ansatz material axioms should give standardized procedures for *constructing* (and classifying) *classes of materials* and constitutive equations belonging to them.

We now formulate

Material Axioms: Constitutive equations have to satisfy:

i) the second law, and if necessary its additionals,

ii) transformation properties by changing the observer,

iii) the material-symmetry.

iv) Additionally state spaces have to be chosen so that hyperbolicity of propagation equations is guaranteed.

In detail the second law is represented by the *dissipation inequality*, changing the observer demands *covariance* in relativistic theories or *objectivity* in non-relativistic theories. Material symmetry is described by *isotropy groups*. Here we are especially interested in i).

### 6.4.1 Exploiting Dissipation Inequality

As an example we choose the 5-field theory, but all items treated in this section are generally valid. The balance equations (6.71) to (6.73) are

$$\dot{\rho} + \rho \underline{\underline{1}} : \nabla \underline{v} = 0 \ , \tag{6.81}$$

$$\rho \dot{\epsilon} + \underline{\underline{1}} : \nabla \underline{q} - \underline{\underline{P}} : \nabla \underline{v} - r = 0 \ , \tag{6.82}$$

$$\rho \dot{\underline{v}} + \nabla \cdot \underline{\underline{P}} - \rho \underline{f} = 0 \ . \tag{6.83}$$

We choose a large state space

$$Z := (\rho, 1/\theta, \nabla\rho, (1/\theta)^{\bullet}, \nabla(1/\theta), \nabla\underline{v}) \tag{6.84}$$

which belongs to an acceleration independent fluid because the acceleration is not included into $Z$. The non-equilibrium variables are

$$\underline{g} := (\nabla\rho, (1/\theta)^{\bullet}, \nabla(1/\theta)) \ , \tag{6.85}$$

and therefore

$$Z = (\rho, 1/\theta, \nabla \underline{v}, \underline{g}) \ . \tag{6.86}$$

The directional derivatives are

$$\underline{y} := (\dot{\underline{g}}, \nabla \underline{g}, \dot{\underline{v}}, (\nabla \underline{v})^{\bullet}, \nabla\nabla \underline{v}) \ . \tag{6.87}$$

Because the balance equations and also the dissipation inequality

$$\rho \dot{s} + \nabla \cdot \underline{\phi} - r/\theta =: \sigma \geq 0 \ , \tag{6.88}$$

$$\underline{\phi} := \underline{k} + \underline{q}/\theta \tag{6.89}$$

are linear in the derivatives $\bullet$ and $\nabla$, we get after having applied the chain rule linear equations in the directional derivatives $\underline{y}$:

$$\underline{\underline{A}} \cdot \underline{y} = \underline{C} \qquad \text{and} \qquad \underline{B} \cdot \underline{y} \geq D \ . \tag{6.90}$$

Because of the chosen large state space: $\underline{\underline{A}}, \underline{C}, \underline{B}, D$ are *state functions* and $\underline{y}$ is beyond the state space variables.

$\underline{\underline{A}}, \underline{C}, \underline{B}, D$ depend on the constitutive equations which are not fixed up to now. $\underline{\underline{A}}, \underline{C}, \underline{B}, D$ contain derivatives of the constitutive equations to the state variables.

*Definition*: All constitutive equations being compatible with the chosen large state space and satisfying the balance equations and the dissipation inequality determine the *class of material*.

In principle there are two possibilities to find this class of material:

i) For fixed $\underline{\underline{A}}, \underline{C}, \underline{B}, D$, the dissipation inequality excludes certain $\underline{y}$, i.e. certain process directions in state space are not allowed.

ii) The $\underline{\underline{A}}, \underline{C}, \underline{B}, D$, have to be determined so that all process directions are possible, i.e. whatever $\underline{y}$ may be, the dissipation inequality is always satisfied.

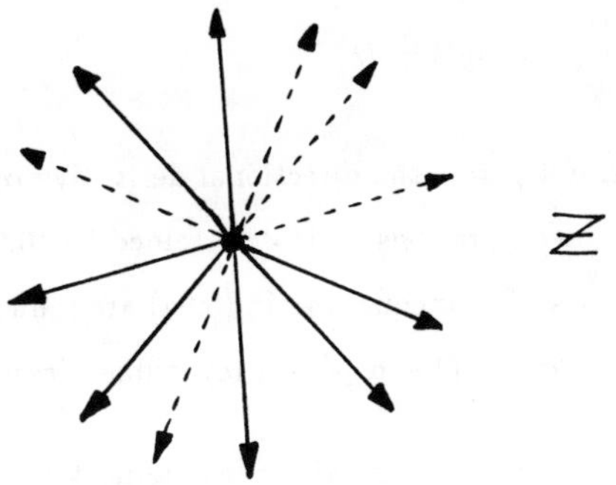


Figure 6.2: At a fixed state $Z$ there are directions represented by the directional derivatives which are allowed ($\rightarrow$) and forbidden ($- - -$). Allowed means in agreement with the dissipation inequality, forbidden in contrast to it.

**Non-Reversible-Direction Axiom**

We consider an arbitrary material at a fixed state $Z(\underline{x}, t)$ [fixed position, fixed time, local]

$$\underline{\underline{A}}(Z), \underline{C}(Z), \underline{B}(Z), D(Z) \ . \tag{6.91}$$

*Assumption*: At $Z$ there are allowed and forbidden directions in state space (Figure 6.2): $\underline{y}^1$ is an *allowed* directional derivative

$$\underline{\underline{A}} \cdot \underline{y}^1 = \underline{C} \ , \qquad \underline{B} \cdot \underline{y}^1 \geq D \ , \tag{6.92}$$

and $\underline{y}^2$ a *forbidden* one

$$\underline{\underline{A}} \cdot \underline{y}^2 = \underline{C} \ , \qquad \underline{B} \cdot \underline{y}^2 < D \ . \tag{6.93}$$

*Proposition*: A consequence of (6.92) and (6.93) is the statement: There exist a reversible direction $\alpha \underline{y}^1 + (1-\alpha)\underline{y}^2$

$$0 < \alpha := \frac{D - \underline{B} \cdot \underline{y}^2}{\underline{B} \cdot (\underline{y}^1 - \underline{y}^2)} < 1 \ . \tag{6.94}$$

*Proof* : Multiplying (6.92) with $\alpha > 0$ and (6.93) with $\beta > 0$ we get by addition:

$$\underline{\underline{A}} \cdot (\alpha \underline{y}^1 + \beta \underline{y}^2) = (\alpha + \beta)\underline{C} \ , \qquad \alpha + \beta \stackrel{!}{=} 1 \ , \tag{6.95}$$

$$\underline{B} \cdot (\alpha \underline{y}^1 + (1-\alpha)\underline{y}^2) \stackrel{!}{=} D \ . \tag{6.96}$$

By the demand $\alpha + \beta = 1$ the directional derivative $\alpha\underline{y}^1 + (1-\alpha)\underline{y}^2$ in (6.95) satisfies the balance equations. $\alpha$ is determined by (6.96) from which immediately (6.94) follows. The inequalities in (6.94) are induced by (6.92) and (6.93). (6.96) shows that $\alpha\underline{y}^1 + (1-\alpha)\underline{y}^2$ is a reversible direction. □

By this proposition we proved the existence of almost one reversible process direction, if at an arbitrary non-equilibrium state $Z$ both kinds of allowed and forbidden directional derivatives exist. By arguments of continuity we conclude the existence not only of a reversible direction but also of a piece of a reversible trajectory. Because we are in non-equilibrium we have to exclude such reversible trajectories:

*Axiom [28]*: Except in equilibrium subspace ***reversible process directions*** in state space *do not exist*.

A consequence of this axiom is the fact that all process directions are either allowed or forbidden, but no non-equilibrium state exists with both kinds of process directions. By the non-reversible-direction axiom and the proposition (6.94) we have proved the

*Proposition*: If $Z$ is no trap, the inclusion

$$\bigwedge \underline{y} : \underline{\underline{A}}(Z) \cdot \underline{y} = \underline{C}(Z) \rightarrow \underline{B}(Z) \cdot \underline{y} \geq D(Z) \tag{6.97}$$

is valid.

By this proposition restrictions of the $\underline{\underline{A}}$, $\underline{C}$, $\underline{B}$, and $D$ result. These restrictions characterize the class of material we are looking for.

## Coleman–Noll Technique

In this technique the inclusion (6.97) is enforced by:

$$\underline{B}(Z) = 0 \wedge D(Z) \leq 0 \tag{6.98}$$

$\underline{\underline{A}}(Z)$ and $\underline{C}(Z)$ are not restricted.

But this choice is only one possibility. The class of materials found by this "technique" is generally too small.

**Liu Technique**

We start out with

*Liu's Proposition [29,30,31]*: By use of the inclusion (6.97) the following statement is valid: Constitutive equations satisfy in large state spaces (1.9) the relations

$$\underline{B}(Z) = \underline{\lambda}(Z) \cdot \underline{\underline{A}}(Z) \ , \tag{6.99}$$

$$\underline{\lambda}(Z) \cdot \underline{C}(Z) \geq D(Z) \ . \tag{6.100}$$

Here the state function $\underline{\lambda}$ is only unique, if $\underline{\underline{A}}$ has its maximal rank. The entropy production density

$$\sigma := \underline{\lambda} \cdot \underline{C} - D \geq 0 \tag{6.101}$$

is independent of directional derivatives.

A corollary results immediately from Liu's proposition.

*Corollary*: For arbitrary directional derivatives $\underline{w}$ which need not satisfy the balance equations we have:

$$\underline{B} \cdot \underline{w} - \underline{\lambda} \cdot (\underline{\underline{A}} \cdot \underline{w} - \underline{C}) \geq D \ . \tag{6.102}$$

**An Additional Proposition**

*Proposition [32]*: We consider a mapping of a vector space $\mathcal{B}$ onto itself

$$\underline{W} : \underline{X} \to \underline{W}(\underline{X}) \ , \qquad \underline{X} \in \mathcal{B} \ , \tag{6.103}$$

If $\underline{W}$ is continuous in the vicinity of $\underline{X} = \underline{0}$, and if

$$\underline{W}(\underline{X}) \cdot \underline{X} > 0 \ , \qquad \text{for all } \underline{X} \ , \tag{6.104}$$

is valid, $\underline{W}$ is homogeneous in its arguments

$$\underline{W}(\underline{X}) = \underline{\underline{a}}(\underline{X}) \cdot \underline{X} \ , \tag{6.105}$$

and $\underline{\underline{a}}$ is positive definite.

We now apply Liu's procedure to a special example.

## 6.4.2 Hyperbolic Heat Conduction

We consider an acceleration independent fluid whose state space is (6.86) and its directional derivatives are (6.87). The balance equations are (6.81) to (6.83), and the dissipation inequality is (6.88).

We now apply the chain rule to the balance of internal energy (6.82) and we get:

$$\begin{aligned} 0 = & -\rho \frac{\partial \epsilon}{\partial \rho} \underline{\underline{1}} : \nabla \underline{v} + \frac{\partial \epsilon}{\partial (1/\theta)} \left(\frac{1}{\theta}\right)^{\bullet} + \rho \frac{\partial \epsilon}{\partial \nabla \underline{v}} : (\nabla v)^{\bullet} \\ & + \rho \frac{\partial \epsilon}{\partial \underline{g}} \cdot \dot{\underline{g}} + \frac{\partial \underline{q}}{\partial \rho} \cdot \nabla \rho + \frac{\partial \underline{q}}{\partial (1/\theta)} \cdot \nabla \frac{1}{\theta} + \frac{\partial \underline{q}}{\partial \nabla \underline{v}} : \nabla \nabla \underline{v} \\ & + \frac{\partial \underline{q}}{\partial \underline{g}} : \nabla \underline{g} + \underline{\underline{P}} : \nabla \underline{v} - r \ . \end{aligned} \tag{6.106}$$

By the same procedure the momentum balance (6.83) results:

$$0 = \rho \dot{\underline{v}} + \frac{\partial \underline{\underline{P}}}{\partial \rho} \cdot \nabla \rho + \frac{\partial \underline{\underline{P}}}{\partial (1/\theta)} \cdot \nabla \frac{1}{\theta} + \frac{\partial \underline{\underline{P}}}{\partial \nabla \underline{v}} : \nabla \nabla \underline{v} + \frac{\partial \underline{\underline{P}}}{\partial \underline{g}} : \nabla \underline{g} - \rho \underline{f} \ . \tag{6.107}$$

The dissipation inequality (6.88) yields:

$$0 \leq \rho \frac{\partial s}{\partial \rho} \dot{\rho} + \rho \frac{\partial s}{\partial (1/\theta)} \left(\frac{1}{\theta}\right)^{\bullet} + \rho \frac{\partial s}{\partial \nabla \underline{v}} : (\nabla v)^{\bullet} +$$

$$+ \rho \frac{\partial s}{\partial \underline{g}} \cdot \dot{\underline{g}} + \left(\frac{\partial \underline{k}}{\partial \rho} + \frac{1}{\theta} \frac{\partial \underline{q}}{\partial \rho}\right) \cdot \nabla \rho + \left\{ \frac{\partial \underline{k}}{\partial (1/\theta)} + \underline{q} + \right.$$

$$\left. + \frac{1}{\theta} \frac{\partial \underline{q}}{\partial (1/\theta)} \right\} \cdot \nabla \frac{1}{\theta} + \left(\frac{\partial \underline{k}}{\partial \nabla \underline{v}} + \frac{1}{\theta} \frac{\partial \underline{q}}{\partial \nabla \underline{v}}\right) \vdots \nabla \nabla \underline{v} +$$

$$+ \left\{ \frac{\partial \underline{k}}{\partial \underline{g}} + \frac{1}{\theta} \frac{\partial \underline{q}}{\partial \underline{g}} \right\} : \nabla \underline{g} - \frac{r}{\theta} \; . \tag{6.108}$$

According to (6.90) we write the balance equations in the form

$$\begin{pmatrix} \underline{\underline{A}} \\ \underline{A} \end{pmatrix} \cdot \underline{y} = \begin{pmatrix} \underline{C} \\ C \end{pmatrix} \tag{6.109}$$

using the abbreviations

$$\underline{\underline{A}} := \left\{ \frac{\partial \underline{\underline{P}}}{\partial \underline{g}}, \underline{0}, \rho, \underline{0}, \frac{\partial \underline{\underline{P}}}{\partial \nabla \underline{v}} \right\} \; , \tag{6.110}$$

$$\underline{A} := \left\{ \frac{\partial \underline{q}}{\partial \underline{g}}, \rho \frac{\partial \epsilon}{\partial \underline{g}}, 0, \rho \frac{\partial \epsilon}{\partial \nabla \underline{v}}, \frac{\partial \underline{q}}{\partial \nabla \underline{v}} \right\} \; , \tag{6.111}$$

$$\underline{C} := - \frac{\partial \underline{\underline{P}}}{\partial \rho} \cdot \nabla \rho - \frac{\partial \underline{\underline{P}}}{\partial (1/\theta)} \cdot \nabla \frac{1}{\theta} - \rho \underline{f} \; , \tag{6.112}$$

$$C := \rho \frac{\partial \epsilon}{\partial \rho} \underline{\underline{1}} : \nabla \underline{v} - \frac{\partial \epsilon}{\partial (1/\theta)} \left(\frac{1}{\theta}\right)^{\bullet} - \frac{\partial \underline{q}}{\partial \rho} \cdot \nabla \rho -$$

$$- \frac{\partial \underline{q}}{\partial (1/\theta)} \cdot \nabla \frac{1}{\theta} - \underline{\underline{P}} : \nabla \underline{v} + r \; . \tag{6.113}$$

The quantities in the dissipation inequality (6.90) are:

$$\underline{B} := \left\{ \frac{\partial \underline{k}}{\partial \underline{g}} + \frac{1}{\theta} \frac{\partial \underline{q}}{\partial \underline{g}}, \rho \frac{\partial s}{\partial \underline{g}}, \underline{0}, \rho \frac{\partial s}{\partial \nabla \underline{v}}, \right.$$

$$\left. \frac{\partial \underline{k}}{\partial \nabla \underline{v}} + \frac{1}{\theta} \frac{\partial \underline{q}}{\partial \nabla \underline{v}} \right\} \; , \tag{6.114}$$

$$D := +\rho\frac{\partial s}{\partial\rho}\underline{\underline{1}} : \nabla\underline{v} - \rho\frac{\partial s}{\partial(1/\theta)}\left(\frac{1}{\theta}\right)^{\bullet}$$

$$-\left(\frac{\partial\underline{k}}{\partial\rho} + \frac{1}{\theta}\frac{\partial\underline{q}}{\partial\rho}\right)\cdot\nabla\rho$$

$$-\left\{\frac{\partial\underline{k}}{\partial(1/\theta)} + \underline{q} + \frac{1}{\theta}\frac{\partial\underline{q}}{\partial(1/\theta)}\right\}\cdot\nabla\frac{1}{\theta} + \frac{r}{\theta} \ . \tag{6.115}$$

The relations (6.99) and (6.100) are in this case

Liu Equations:

$$\underline{\underline{B}} = \underline{\lambda}\cdot\underline{\underline{A}} + \lambda\underline{A} \ , \tag{6.116}$$

Reduced Dissipation Inequality

$$\underline{\lambda}\cdot\underline{C} + \lambda C \geq D \ . \tag{6.117}$$

The Liu equations are explicitly:

$$\frac{\partial\underline{k}}{\partial\underline{g}} + \frac{1}{\theta}\frac{\partial\underline{q}}{\partial\underline{g}} = \underline{\lambda}\cdot\frac{\partial\underline{\underline{P}}}{\partial\underline{g}} + \lambda\frac{\partial\underline{q}}{\partial\underline{g}} \tag{6.118}$$

$$\rho\frac{\partial s}{\partial\underline{g}} = \lambda\frac{\partial\epsilon}{\partial\underline{g}} \tag{6.119}$$

$$\underline{0} = \underline{\lambda}\rho \ \rightarrow \ \underline{\lambda} = \underline{0} \ . \tag{6.120}$$

$$\rho\frac{\partial s}{\partial\nabla\underline{v}} = \lambda\rho\frac{\partial\epsilon}{\partial\nabla\underline{v}} \tag{6.121}$$

$$\frac{\partial\underline{k}}{\partial\nabla\underline{v}} + \frac{1}{\theta}\frac{\partial\underline{q}}{\partial\nabla\underline{v}} = \underline{\lambda}\cdot\frac{\partial\underline{\underline{P}}}{\partial\nabla\underline{v}} + \lambda\frac{\partial\underline{q}}{\partial\nabla\underline{v}} \tag{6.122}$$

The reduced dissipation inequality is:

$$\left\{-\lambda\underline{\underline{P}} - \rho\left(\frac{\partial s}{\partial\rho} - \lambda\frac{\partial\epsilon}{\partial\rho}\right)\underline{\underline{1}}\right\} : \nabla\underline{v}$$

$$+\rho\left\{\frac{\partial s}{\partial(1/\theta)} - \lambda\frac{\partial\epsilon}{\partial(1/\theta)}\right\}\left(\frac{1}{\theta}\right)^{\bullet}$$

$$+\left\{\frac{\partial\underline{k}}{\partial\rho} + \left(\frac{1}{\theta} - \lambda\right)\frac{\partial\underline{q}}{\partial\rho}\right\}\cdot\nabla\rho \ +$$

$$+\left\{\underline{q}+\frac{\partial\underline{k}}{\partial(1/\theta)}+\left(\frac{1}{\theta}-\lambda\right)\frac{\partial\underline{q}}{\partial(1/\theta)}\right\}\cdot\nabla\frac{1}{\theta}$$

$$+\left\{\lambda-\frac{1}{\theta}\right\}r\geq 0\ . \tag{6.123}$$

Without taking into consideration external forces the *equilibrium conditions* are

$$\nabla\underline{v}=\underline{\underline{0}}\ ,\qquad (1/\theta)^{\bullet}=0\ , \tag{6.124}$$

$$\nabla\rho=0\ ,\qquad \nabla(1/\theta)=\underline{0}\ , \tag{6.125}$$

$$\underline{q}=\underline{0}\ ,\qquad \sigma=0\ . \tag{6.126}$$

Therefore from (6.123) results in equilibrium

$$\lambda|_E=1/\theta\ . \tag{6.127}$$

Applying (6.105) to (6.123) we get

$$\underline{\underline{P}}|_E=P\underline{\underline{1}}\ ,\qquad P:=-\theta\rho\left\{\left.\frac{\partial s}{\partial\rho}\right|_E-\frac{1}{\theta}\left.\frac{\partial\epsilon}{\partial\rho}\right|_E\right\} \tag{6.128}$$

$$\left.\frac{\partial s}{\partial(1/\theta)}\right|_E=\frac{1}{\theta}\left.\frac{\partial\epsilon}{\partial(1/\theta)}\right|_E\ ,\qquad \left.\frac{\partial\underline{k}}{\partial\rho}\right|_E=\underline{0}\ ,\qquad \left.\frac{\partial\underline{k}}{\partial(1/\theta)}\right|_E=\underline{0}\ . \tag{6.129}$$

From (6.127) we get in non-equilibrium:

$$\lambda-\frac{1}{\theta}=\underline{\underline{\ell}}^1:\nabla\underline{v}+\ell^2\left(\frac{1}{\theta}\right)^{\bullet}+\underline{\ell}^3\cdot\nabla\rho+\underline{\ell}^4\cdot\nabla\frac{1}{\theta}\ . \tag{6.130}$$

Here the $\ell^j$ are state functions. From (6.130) results:

$$\left.\frac{\partial\lambda}{\partial(1/\theta)}\right|_E=1\ ,\qquad \left.\frac{\partial^n\lambda}{\partial(1/\theta)^n}\right|_E=0\ ,\qquad n\geq 2\ , \tag{6.131}$$

$$\left.\frac{\partial^m\lambda}{\partial\rho^m}\right|_E=0\ ,\qquad m\geq 1\ . \tag{6.132}$$

We now consider the special case of *pure heat conduction* which is characterized by

$$\nabla\underline{v}=\underline{\underline{0}}\ ,\qquad \nabla\rho=\underline{0}\ ,\qquad r=0\ . \tag{6.133}$$

From (6.123) and from (6.130) we get:

$$\rho\left\{\frac{\partial s}{\partial(1/\theta)}-\lambda\frac{\partial \epsilon}{\partial(1/\theta)}\right\}=\hat{\alpha}\left(\frac{1}{\theta}\right)^{\bullet}+\underline{\hat{\beta}}\cdot\nabla\frac{1}{\theta}\ , \tag{6.134}$$

$$\underline{q}+\frac{\partial \underline{k}}{\partial(1/\theta)}+\left(\frac{1}{\theta}-\lambda\right)\frac{\partial \underline{q}}{\partial(1/\theta)}=\underline{\hat{a}}\left(\frac{1}{\theta}\right)^{\bullet}+\underline{\underline{\hat{b}}}\cdot\nabla\frac{1}{\theta}\ , \tag{6.135}$$

$$\lambda-\frac{1}{\theta}=\ell^{2}\left(\frac{1}{\theta}\right)^{\bullet}+\underline{\ell}^{4}\cdot\nabla\frac{1}{\theta}\ . \tag{6.136}$$

From (6.106) we get the *linearized* balance equation of internal energy in the case of pure heat conduction

$$\rho\left.\frac{\partial \epsilon}{\partial(1/\theta)^{\bullet}}\right|_{E}\left(\frac{1}{\theta}\right)^{\bullet}+\rho\left.\frac{\partial \epsilon}{\partial\nabla(1/\theta)^{\bullet}}\right|_{E}\cdot\left(\frac{1}{\theta}\right)^{\bullet\bullet}+\rho\left.\frac{\partial \epsilon}{\partial\nabla(1/\theta)}\right|_{E}\left(\nabla\frac{1}{\theta}\right)^{\bullet}$$

$$=-\nabla\cdot\underline{q}$$

$$=-\nabla\cdot\left\{-\frac{\partial \underline{k}}{\partial(1/\theta)}+\left(\underline{\hat{a}}+\ell^{2}\frac{\partial \underline{q}}{\partial(1/\theta)}\right)\left(\frac{1}{\theta}\right)^{\bullet}+\right.$$

$$\left.+\left(\underline{\underline{\hat{b}}}+\underline{\ell}^{4}\frac{\partial \underline{q}}{\partial(1/\theta)}\right)\cdot\nabla\frac{1}{\theta}\right\}\ . \tag{6.137}$$

From (6.134) we have

$$\rho\frac{\partial}{\partial(1/\theta)}(-\lambda\psi)+\rho\frac{\partial \lambda}{\partial(1/\theta)}\epsilon=\hat{\alpha}\left(\frac{1}{\theta}\right)^{\bullet}+\underline{\hat{\beta}}\cdot\nabla\frac{1}{\theta}\ , \tag{6.138}$$

by introducing the free energy $\psi$

$$\lambda\psi:=\lambda\epsilon-s\ . \tag{6.139}$$

We now differentiate (6.138) and consider the equilibrium:

$$\left.\frac{\partial}{\partial(1/\theta)^{\bullet}}\right|_{E}\ ,\qquad\left.\frac{\partial}{\partial\nabla(1/\theta)}\right|_{E}\ . \tag{6.140}$$

Then we get:

$$-\rho\left.\frac{\partial}{\partial(1/\theta)}\frac{\partial}{\partial(1/\theta)^{\bullet}}\lambda\psi\right|_{E}+\rho\left.\frac{\partial \lambda}{\partial(1/\theta)^{\bullet}\partial(1/\theta)}\right|_{E}\epsilon+$$

$$+\rho\left.\frac{\partial \lambda}{\partial(1/\theta)}\right|_{E}\left.\frac{\partial \epsilon}{\partial(1/\theta)^{\bullet}}\right|_{E}=\hat{\alpha}|_{E}\ . \tag{6.141}$$

The second differentiation in (6.140) yields a similar equation in which $(1/\theta)^\bullet$ is replaced by $\nabla(1/\theta)$ and the right-hand side is $\underline{\hat{\beta}}|_E$. Introducing the abbreviations

$$\Psi^1 = \frac{\partial}{\partial(1/\theta)}\frac{\partial}{\partial(1/\theta)^\bullet}\lambda\psi|_E \ , \tag{6.142}$$

$$\underline{\Psi}^2 = \frac{\partial}{\partial(1/\theta)}\frac{\partial}{\partial\nabla(1/\theta)}\lambda\psi|_E \ , \tag{6.143}$$

and using (6.136) we get a system of *linear* and *inhomogeneous* heat conduction equations of *second order* in $1/\theta$. When we put the variables into the following order

$$(1/\theta)^{\bullet\bullet} \ , \qquad (\nabla(1/\theta))^\bullet \ , \qquad \nabla(1/\theta)^\bullet \ , \qquad \nabla\nabla(1/\theta) \ ,$$

we have for the matrix of the leading terms which determines the type of the heat conduction equation:

$$\begin{bmatrix} \hat{\alpha}|_E - \rho\left.\dfrac{\partial \ell^2}{\partial(1/\theta)}\right|_E \epsilon + \rho\Psi^1 & \underline{\hat{\beta}}|_E - \rho\left.\dfrac{\partial \underline{\ell}^4}{\partial(1/\theta)}\right|_E \epsilon + \rho\underline{\Psi}^2 \\ \underline{\hat{a}}|_E + \ell^2\left.\dfrac{\partial \underline{q}}{\partial(1/\theta)}\right|_E & \underline{\underline{\hat{b}}}|_E + \underline{\ell}^4\dfrac{\partial \underline{q}}{\partial(1/\theta)} \end{bmatrix} \tag{6.144}$$

From (6.123) we get by use of (6.134) and (6.135) the bilinear form of the entropy production density:

$$\hat{\alpha}\left[\left(\frac{1}{\theta}\right)^\bullet\right]^2 + \underline{\hat{\beta}}\cdot\nabla\frac{1}{\theta}\left(\frac{1}{\theta}\right)^\bullet + \underline{\hat{a}}\left(\frac{1}{\theta}\right)^\bullet\cdot\nabla\frac{1}{\theta} + \underline{\underline{\hat{b}}}:\nabla\frac{1}{\theta}\nabla\frac{1}{\theta} \geq 0 \ . \tag{6.145}$$

Therefore the matrix of this bilinear form in equilibrium is positive definite:

$$\begin{bmatrix} \hat{\alpha}|_E & \underline{\hat{\beta}}|_E \\ \underline{\hat{a}}|_E & \underline{\underline{\hat{b}}}|_E \end{bmatrix} \geq 0 \ . \tag{6.146}$$

A comparison with (6.144) shows that even if

$$\lambda \equiv 1/\theta \rightarrow \ell^2 \equiv 0 \ , \quad \underline{\ell}^4 \equiv \underline{0} \ , \tag{6.147}$$

is valid, we have additional terms $\Psi^1$ and $\underline{\Psi}^2$ in (6.144) which induce hyperbolicity. Therefore hyperbolic heat conduction is also possible in 5-field theories, if the state space is chosen appropriately.

In this example the material axioms ii) and iii) in Section 6.4 have now to be taken into consideration. But this is out of scope of these lectures.

# References


1. Schottky, W., Thermodynamik, 1. Teil § 1, Springer-Verlag, Berlin (1929).

2. Axelrad, D.R.: Foundations of the Probabilistic Mechanics of Discrete Media. Oxford: Pergamon Press 1984.

3. Muschik, W.: Thermodynamical Constitutive Laws — Outlines, in Axelrad, D.R. and Muschik, W. (Eds.): Constitutive Laws and Microstructures, Springer, Berlin 1988.

4. Fick, E.; Sauermann, G.: Quantenstatistik Dynamischer Systeme, Bd. I. Thun: Harri Deutsch 1983.

5. Cercignani, C.: Theory and Application of the Boltzmann Equation. New York: Elsevier 1975.

6. Eu, B.C.: J. Chem. Phys. 87 (1987) 1220.

7. Meixner, J.: Rheol. Acta 7 (1968) 8.

8. Muschik, W.: J. Non-Equilib. Thermodyn. 4 (1979) 277.

9. Muschik, W.: J. Appl. Sci. 4 (1986) 189.

10. Muschik, W.: Arch. Rat. Mech. Anal. 66 (1977) 379.

11. Purcell, E.M.; Pound, R.V.: Phys. Rev. 81 (1951) 279.

12. Proctor, W.G.: Scient. American 239 (1978) 90.

13. Muschik, W.: Formulations of the Second Law, Symposium "Statistical Thermodynamics and Semiconductors" Gregynog, Sept. 14–17, 1987, J. Phys. Chem. Solids 49 (1988) 709

14. Kestin, J.: A Course in Thermodynamics, Vol. 1, § 9.8. McGraw-Hill, New York (1979).

15. Muschik, W.: Int. J. Engng. Sci. 18 (1980) 1399.

16. Muschik, W.: J. Non-Equilib. Thermodyn. 4 (1979) 377.

17. Muschik, W.; Brunk, G.: Int. J. Engng. Sci. 15 (1977) 377.

18. Muschik, W.: Continuum Models of Discrete Systems 4, Brulin, O. and Hsieh, R.K.T. (Eds.), p.511, North-Holland, Amsterdam (1981).

19. Fick, E.; Schwegler, H.: Z. Physik 200 (1967) 165.

20. Jaynes, E.T.: Phys. Rev. 106 (1957) 620, 108 (1957) 171.

21. Robertson, B.: Phys. Rev. 144 (1966) 151.

22. Kawasaki, K.; Gunton, J.D.: Phys. Rev. A8 (1973) 2048.

23. Muschik, W.: in Spencer, A.J.M. (ed.): Continuum Models of Disctete Systems. Rotterdam: Balkema 1987, p.39.

24. Muschik, W.: J. Non-Equilib. Thermodyn. 8 (1983) 219.

25. Müller, W.H.; Muschik, W.: J. Non-Equilib. Thermodyn. 8 (1983) 29, 47.

26. Müller, I.: in Lecture Notes in Physics 199. Berlin: Springer 1984, p.32.

27. Liu, I.S.; Müller, I.: Arch. Rat. Mech. Anal. 83 (1983) 285.

28. Muschik, W.: in Disequilibrium and Self-Organisation, Kilmister, C.W. (Ed.), Reidel 1986, p.65.

29. Liu, I.S.: Arch. Rat. Mech. Anal. 46 (1972) 131.

30. Muschik, W.: in Lecture Notes in Physics 199. Berlin: Springer 1984, p.387.

31. Muschik, W.; Ellinghaus, R.: ZAMM 68 (1988) T232.

32. Muschik, W.: in Chien Wei-zang (ed.): Proc. Intern. Conf. Nonlinear Mech. (ICNM) Shanghai 1985, p.155.